SHRUB ROSES

in Australia and New Zealand

Also by Deane Ross
GROWING ROSES

SHRUB ROSES

in Australia
and New Zealand

DEANE ROSS

Maiden's Blush

Rigby

RIGBY LIMITED • ADELAIDE • SYDNEY
MELBOURNE • BRISBANE • PERTH

First published 1972
Copyright © 1972 Deane Ross
Library of Congress Catalog Card Number 71-160569
National Library of Australia Card Number
and ISBN 0 85179 384 3
All rights reserved
Wholly designed and set up in Australia

Printed in Hong Kong

CONTENTS

LIST OF ILLUSTRATIONS

AUTHOR'S NOTE

I have chosen to consider shrub roses in nine groups of similar styles and usages, consistent with close botanic grouping. Most varieties fall conveniently into these categories, but inevitably some could go into two or more groups while others fail logically to fit into any.

Each group or section is prefaced by information about their history, origin, characteristics, uses, culture, and pruning. In addition, information concerning their typical growth, flowers, and usage saves repetition in subsequent descriptions, so that their most important features are mentioned in the group description. The accompanying photographs are not intended to favour or recommend any particular variety, but rather to show the typical form of that group or class.

The first or main names are the most accurate that I can ascertain, followed by any synonymous and common names in parentheses ( ). Species are listed alphabetically in each section, omitting the genus *Rosa*. For convenience, synonyms are cross-indexed in the general index.

In order to avoid unnecessary repetition in the descriptions and to provide a quick reference of important features, the following codes have been adopted:

Height:
1. very low;
2. low;
3. medium;
4. tall;
5. particularly tall.

Spread:
a. very upright in growth;
b. upright;
c. spreading;
d. very spreading.

Blooms:
    s. single or five petals;
    ss. semi-single (barely two rows of petals);
    sd. semi-double (usually up to about twenty-five petals);
    d. double (maximum number of petals).

Flowering Period:
    S. spring only;
    R. recurrent, some autumn blooms;
    P. perpetual, one flowering follows another.

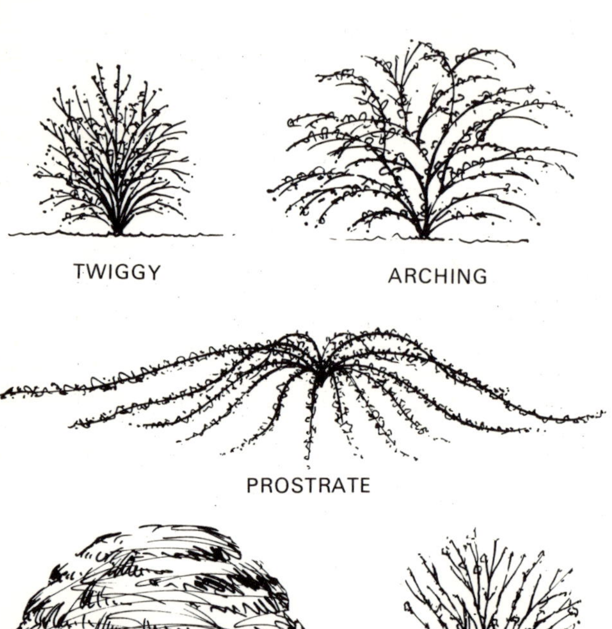

TWIGGY   ARCHING

PROSTRATE

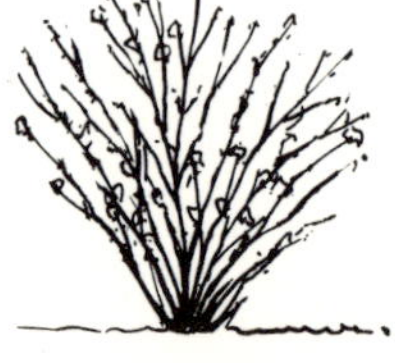

CASCADING   THICKETING

INTRODUCTION

It is quite certain that whoever picks up this book and reads the word "Roses" in the title will have a clear picture of a rose in his mind. This will be of a large classically shaped Hybrid Tea rose, as this indeed is the ultimate creation of the modern breeder and the most popular concept of the modern rose.

However, upon reading the title "Shrub Roses" the mental picture may become blurred, for few people ever think of roses as shrubs. The individual rose blooms take our attention and scant thought is given to the form and habit of the plant that bears it. Of course, by strict definition, all roses are shrubs, whether modern Hybrid Teas, Floribundas, the species, or cultivated varieties. However since the field of popular Hybrid Teas, Floribundas, Miniatures, and Climbers have been adequately covered, this book omits these and discusses the lesser known but nevertheless valuable garden roses under the all-embracing term of "Shrub Roses."

The purpose of this book is threefold. Firstly, it aims to introduce the reader to shrub roses, with the hope that it will open up a wider field of gardening enjoyment to him. The interest in shrub roses in Australia has heightened in the past few years. This could be because so many people are travelling overseas and seeing them in the "old" countries where they are more widely grown, or it may be because of several magazine articles published recently.

Secondly, although several good books have been written overseas on shrub roses, there is a need for a volume to cover them under Australian conditions, especially regarding the culture, flowering times, and those varieties that are reasonably available commercially. Lastly, it aims to smooth the way (if necessary) for intending growers with cultural advice, pruning,

Chapeau de Napoléon

and training, together with recommended siting of the various types and varieties.

The story starts with the original species or wild roses and their close hybrids, through the early cultivated roses to the many and varied developments of the 1800s, and finally to the more recent shrub roses which, because they do not conform to the popular concept of roses, have been neglected by most rose growers. Some of these display exceptional virtues of colour, vigour, hardiness, and floriferousness, others are grown because of their historic or botanic significance. Among them will be found roses with the most diverse features—large or small blooms, fully double to simple five petals, with colours extending to wider extremes than the moderns; some with ornamental hips or seed pods; some petite, others brutishly vigorous; and practically all suitable for making a contribution to the garden.

It seems to me a quaint paradox that despite intensive and prolific breeding by modern hybridists, the form of the rose flower and the size and shape of the plant differ only slightly from one another, with nowhere near the interesting variations that are found in shrub roses. In fact, colour alone seems to distinguish some new roses from others. Breeders try to please the buying public and if the public showed a greater interest in shrub roses, breeders would soon respond to the demand. It is a pity that such stereotype thinking should dominate the present development of the rose.

I also believe that too much emphasis is placed on the perfection of the individual bloom as distinct from the shrub effect of the blooms in relation to their plant. Competitive rose shows consider only the specimen blooms taken out of their context with the plant, as also do photographs of individual blooms that are frequently published in magazines and catalogues. Thus many rose growers choose the varieties for their gardens by the bloom alone. To offset this, I have endeavoured to consider each shrub rose as a unit in the landscape. Each has its particular habits, and no rose is condemned outright just because it does not fit into a preconceived situation. Large or small, open or dense, spreading or upright, all have a place if you understand their characteristics. It is possible to landscape an entire garden with roses without monotony.

Mme Hardy

I have lived and worked among the shrubs for many years now, but it was not until I commenced the detailed preparation for this book that I realised how much research has been done to bring some semblance of order out of the conflict and confusion of early records and illustrations. Use has been made of *Modern Roses 7*, the official reference volume of the International Registration Authority for Roses. I have also gleaned

La Reine Victoria

information from the writings of Dr C. C. Hurst and Mr G. S. Thomas of England.

Many of the old shrub roses were brought into Australia by the earlier settlers and have persisted to this day, although the names handed down by word of mouth are often in error. Some names are difficult to identify while others are rather suspect.

On the other hand, many of the shrub roses that are available at present have been newly imported into Australia from reliable overseas sources, and it is to be hoped that not only may they be preserved in appropriate gardens but that their names may remain accurate.

Finally, I unashamedly commend shrub roses to you when I suggest that you will take a step back into the past. In these days with "progress" rushing ahead, when one does not have time to try one variety before another comes that is bigger, better, and brighter, when you must "keep up with the Joneses" even in the rose garden, the shrub roses, especially the old world roses, will help you to pause for a moment. Their muted colours suggest an age before neon signs, their ruffled blooms suggest the swaying crinoline, their five petals suggest the bushland wild flowers, and their simple pure fragrance is a rare breath of unpolluted country air. From the simple wild roses and the favourites of an earlier century you will discover a new dimension in gardening pleasure.

SPECIES ROSES—WHERE IT ALL BEGAN

We know that cultivated plants are developed, either by chance or plan, from certain of their wild or species parents. This is equally true of the rose, which has a particularly rich selection of species material dispersed throughout the northern hemisphere. Over a hundred species and a similar number of natural hybrids have been found in habitats extending from Alaska to Mexico, from Siberia to Abyssinia, throughout the continents of the northern hemisphere. Each species has adapted to the rigours of its own environment, whether snowy landscape, dry desert, tropical forest, or temperate plain.

Botanically, the genus *Rosa* belongs to the family Rosaceae, and is "first cousin" to such well known commercial and garden plants as *Prunus* (plums), *Malus* (apples), *Rubus* (berries), *Cotoneaster*, *Spiraea*, and *Sorbus* (rowan). Despite the wide range of species available, surprisingly few have been used in modern breeding. Other species are either incompatible for breeding with our moderns, or else their virtues are overwhelmed by some dominant disadvantage, such as excessive thorniness or susceptibility to disease.

For botanical convenience, the genus *Rosa* has been grouped into sections or sub-genera, and it is logical and of interest to consider them in this way. The sub-genera used in modern breeding and their contribution to the present position are:

Indicae (the China Rose) revolutionised the nineteenth century rose world because of their perpetual flowering ability. The tea scent, the loose nodding blooms, and the fine twiggy habit also come through the China roses, whose main colours are soft yellow, pink, and crimson.

Gallicanae strongly influenced the Old European roses, and gave us the double blooms of dusky pink, red, and purple, with rich damask perfume. The growth is thick and upright, particularly frost and blackspot resistant, although at times subject to mildew.

Caninae (the Dog Rose) are usually of simple five-petalled form in shades of soft cream and pink, while the foliage is soft green and matt, characteristics that are evident in the Alba roses.

Synstylae are climbers, three of which are worthy of special note. *Moschata* (Musk Rose) has had a very significant effect throughout rose history. It gave the first tendency towards recurrent flowering in the Old European roses. Later it was used in hardy, fragrant shrub roses, and more recently still, in the Floribunda class. *Wichuraiana* is the vigorous but completely prostrate rambler which produced the "Dorothy Perkins" type ramblers, besides being used in the more orthodox climbers. *Multiflora*, with its huge trusses of bloom, helped to give our Floribundas the large heads of bloom which are eagerly sought after for floral use.

Pimpinellifoliae, as the name implies, have fine fern-like leaves, and as such, make quite ornamental shrubs. Most flowers are of single form, yellow to white or pale pink, flowering quite early in the spring. They provide, through the heavily scented *foetida*, the most valuable yellow breeding ingredient in modern roses. Others that are typical of the group are *spinosissima* (Burnet Rose), and *hugonis*.

Cinnamomeae are a large, rather diverse group named after the Cinnamon Rose, an otherwise rather insignificant rose. They have not been used to any extent in breeding, but most are fine looking garden shrubs, especially noted for interesting seed hips and bright small single flowers. Although most, such as *moyesii*, have an open arching habit of growth, *rugosa* is in direct contrast with low, very dense, and lush bushes. Colours in this group range from purple through cinnamon-pink to white.

Several species are quite distinct from the main stream of the above roses, almost to the point of being regarded as not

DEVELOPMENT OF THE MODERN ROSE

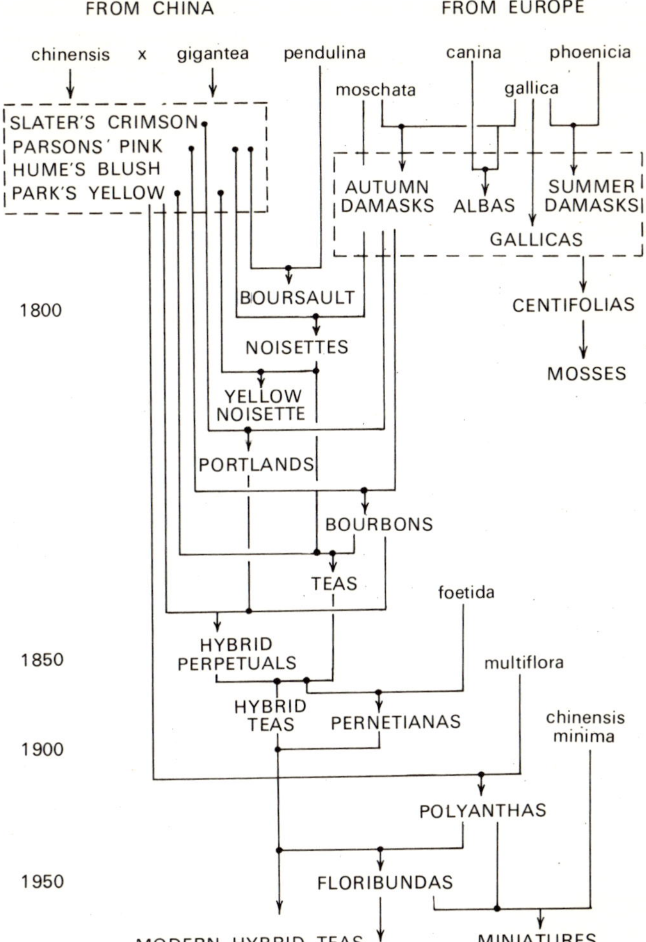

true *Rosa. Banksiae* (Yellow and White Banksias) are the best known of these, also *laevigata* (Cherokee Rose), *roxburghii* (Burr Rose), and *stellata*. These are also valuable garden plants, bearing in mind their particular habits of growth and form. Most have defied all efforts to be bred into the modern roses.

SPECIES AND THEIR HYBRIDS

THIS group brings together a widely diverse selection of roses. In order to make the choice of roses for specific purposes easier, the climbers have been grouped separately, while some of the more recent species hybrids which differ markedly from their species parent are grouped under Modern Shrubs.

As it is not possible to generalise upon the habits and usage of such a diverse group, each individual description aims to cover these matters. However, while it is useful to remember that wild or species roses are the produce of natural selection and evolution, they are extremely hardy and disease resistant, especially when grown in similar conditions to their habitat.

Pruning consists merely of shaping the bush when small, removing old crowded branches, and tidying up the mature bushes. The flowering habits of each particular species should be used as a guide, since the species differ widely. Some flower close along the second year's canes, so that winter trimming of these canes should be minimal. Where hips are the feature, trim in order to maintain the best display. Where stems, foliage, or thorns are features, such as for floral art use, prune harder in winter to encourage strong spring growth.

AGNES (*rugosa* × *foetida persiana*)
The compact pompon blooms of light amber-yellow show an obvious *foetida* influence, while the small light green foliage and upright habit combine the features of both parents. The blooms are sharply scented. It was introduced in 1922.   3b. d. R.

BLANDA (LABRADOR ROSE)
Introduced from America in 1773, it has single, soft white blooms, two inches in diameter, with cream stamens. They

develop in small clusters on a clean open bush, later setting globular hips.   4c. s. S.

CANARY BIRD (possibly *hugonis* × *xanthina*)
Noted for its large, single, bright yellow flowers, the plant is large and twiggy with fern-like foliage of rich green.   4c. s. S.

CANTABRIGIENSIS (or *pteragonis cantabrigiensis*)
A recent hybrid between *hugonis* and *sericea*, it makes a particularly attractive shrub. It is of thicketing habit, having long canes covered with a fur of harmless prickles. The foliage is fern-like, turning bronze in autumn, while the small soft yellow blooms are closely packed along each cane. An excellent rose for floral arrangements.   3c. s. S.

CINNAMOMEA PLENA (Double Cinnamon Rose, Rose de Mai, Stevens Rose)
Cultivated prior to 1600, it was a popular early species, but now is mainly of botanic and historic interest. It has a small, crinkled double form of lilac-pink, and flowers early in spring. 2c. sd. S.

DAVIDII
Introduced in 1908, it features clusters of single lilac-pink flowers on an angular bush of smooth cinnamon wood, with sparse leaves and attractive long hips.   4b. s. S.

DUPONTII (probably *moschata* × *gallica*)
This is a large, loose shrub which is significant for its massive display of blooms and hips. Large loose sprays of airy small creamy-white blooms, banana scented, are followed by equally airy sprays of small orange hips. It was introduced prior to 1817.   5c. s. S.

ECAE
This species was introduced in 1880 from Afghanistan. It has tall but twiggy growth of very deep bronze and fine foliage, giving a curiously airy effect. The small single flowers are the most intense buttercup yellow.   4b. s. S.

EGLANTERIA (*rubiginosa*, Eglantine, Sweet Brier Rose)
This shrub is worth growing for the fragrance of the foliage alone, which pervades the whole garden on a warm still day. The small, single, pale pink blooms are followed by small,

oval, orange hips on an upright plant with light green aromatic foliage.   4b. s. S.

Eglanteria LADY PENZANCE (*penzançeana, eglanteria* × *foetida bicolor*)
The one and a half inch diameter single blooms are soft yellow, delicately edged with pink, and with golden stamens. The dense, small, deep green foliage is sharply scented like *eglanteria*.   4b. s. S.

FARRERI PERSETOSA
The rare Threepenny-bit Rose, noted for its tiny lilac-pink flowers and fine ferny foliage, which make a curious contrast with the thick mass of hairy thorns on the stems.   3c. s. S.

FEDTSCHENKOANA
Introduced in 1876 from Turkestan. This fine species shrub should be in all but the smallest gardens. The grey foliage with amethyst tinted tips forms a dense, well rounded mass to eight feet high. The two inch diameter single white blooms are scattered over the bush for a very long period, followed by a light crop of deep red hips.   5c. s. R.

FOETIDA BICOLOR (Austrian Copper)
This species was recorded prior to 1590, and is the best known of the three *foetidas*. All three have similar habits, with twiggy bushes of smooth brown wood and bright green ferny foliage. They flower for an extended spring season and with occasional blooms into autumn. The scent, from which they get their name, is pungent and characteristic. *Bicolor* has a single bloom, up to three inches in diameter, nasturtium red in colour with warm yellow on the reverse and buds.   3c. s. S.

FOETIDA LUTEA (Austrian Brier, Austrian Yellow)
The bright yellow blooms with matching stamens give a particularly intense effect to this shrub.   3c. s. S.

FOETIDA PERSIANA (Persian Yellow)
The double form of the *lutea*, reputed to be the parent of the Pernetiana class, and subsequently of our modern bright yellows, orange, and bicolours.   3c. d. S.

HARDII

This shrub is botanically interesting as a unique hybrid between a *Rosa* and *Hulthemia*, a closely related genus. It is a satisfactory garden shrub with fine twiggy growth and very slim leaves. The two inch diameter single flowers are yellow with a large red "eye" at the base of each petal, and are borne over an extended spring and summer season.   3c. s. R.

HARISONII (HARISON'S YELLOW ROSE, possibly *foetida* × *spinosissima*)

Harisonii resembles the Persian Yellow, but the clear yellow blooms are slightly cupped, while the bush is quite vigorous and upright.   4b. sd. S.

HUGONIS (FATHER HUGO ROSE)

Introduced from China prior to 1900, this is a charming shrub, with elegant arching canes and very fine foliage. The slightly cupped blooms, two inches in diameter, are soft yellow, and are packed along the maroon branches.   4c. s. S.

MACRANTHA

This species and its hybrids produce some of the showiest single blooms. *Macrantha* has large blush-pink blooms with prominent apricot stamens, and is followed by large round hips. The bush is strong, but open and arching.   4d. s. S.

Macrantha COMPLICATA (hybrid)

This is most likely a hybrid, *macrantha* × *gallica*, favouring the former parent. It has the largest most perfectly formed single blooms imaginable, clear rose-pink in colour and paling toward the soft gold stamens, with a strong fruity fragrance. It flowers for a short concentrated period on a large sprawling plant.   3d. s. S.

Macrantha DÜSTERLOHE

The large single crinkled blooms of rose-pink with yellow stamens develop on a large spreading thorny bush.   4d. s. S.

Macrantha LADY CURZON (*macrantha* × *rugosa*)

A strong sprawling shrub, very thorny, this is best suited to large gardens. However, the large crinkled blooms of soft lilac-pink with yellow stamens flower *en masse* with striking effect.   5c. s. S.

Macrantha RAUBRITTER

Unlike other *macranthas*, the plant is low, spreading and twiggy, making it ideal for low front plantings. The striking feature is the nodding, globular shape of the semi-double pink blooms, useful for the drooping extremities of "old world" floral arrangements.   1d. sd. S.

MOYESII

Introduced from China about 1900, *moyesii* and its descendants brought a new brilliance into the species garden. Masses of bright crimson blooms, small and single, festoon each of the arching canes late in spring. These are followed by a crop of hips which are the most striking of any species. The large orange hips, long and flagon-shaped with persistent sepals, hang until the following season. The plant is open and arching with a fine leaf pattern, while the blossoms have a fruity fragrance. As these types flower on mature wood, they should be pruned lightly after flowering.   4c. s. S.

MOYESII FARGESII

A variant of *moyesii* with cerise-pink flowers and sealing-wax hips. It is the most likely parent of later hybrids, being tetraploid.   4c. s. S.

Moyesii EOS

A particularly vigorous *moyesii* hybrid with masses of semi-single coral-red blooms with yellow stamens. This hybrid does not set hips.   5c. ss. S.

Moyesii GERANIUM

A *moyesii* variety with bright crimson blooms and orange hips. However it is more tidy and compact than other varieties, making it a natural choice in smaller gardens.   3c. s. S.

MOYESII HIGHDOWNENSIS

Another *moyesii* variety with blooms of cerise-crimson, and large orange-scarlet hips. This would be the first choice in this group, especially where space is adequate.   4c. s. S.

NITIDA

Introduced prior to 1807, this shrub is distinctly *rugosa* like, with bright lilac-pink single blooms and yellow stamens, later setting round red hips. Thick glossy green foliage turns red in autumn. It is low and thick and is ideal for front row planting. 1c. s. S.

POMIFERA DUPLEX (WOLLEY DOD'S ROSE, APPLE-SCENTED ROSE)

The pretty yet small clear pink, semi-single flowers have yellow stamens, and are borne close along the stout arching branches. Interesting features are the bristles on both the buds and the large crimson hips, and also the apple-like fragrance. 4c. ss. S.

PRIMULA (INCENSE ROSE)

This is so named because the young tips, when bruised, give off an incense-like perfume. The plant, with its upright thicketing canes closely set with soft primrose flowers and deep green ferny foliage, is a worthy garden shrub.   3b. s. S.

ROXBURGHII PLENA (*microphylla plena*, CHESTNUT ROSE, BURR ROSE)

Introduced from China in 1814, it has several botanical features unusual to *Rosa*: bark which peels upon maturing, pairs of ferocious upward facing prickles, and prickly chestnut-like buds. The double blooms are borne over a long spring season, and are a pale pink, shading deeper toward the centre, while the green hips are also prickly. The bush is densely covered with fine ferny foliage.   4d. d. S.

RUBRIFOLIA

A large open bush, and as the name suggests, the stems and foliage of this striking species are an attractive reddish-brown. The same theme is carried through the small single carmine-pink blooms and the red-brown hips, making it a useful and interesting shrub.   4c. s. S.

Rubrifolia CARMENETTA (*rubrifolia × rugosa*)

More closely resembling the *rubrifolia* parent, being only slightly less red. The spring blooms are delightful—simple rose-pink singles, with long spider-like sepals alternating between the slim petals. It flowers in large clusters and sets red hips.   4c. s. S.

SERICEA PTERACANTHA (*omeiensis pteracantha*, MALTESE CROSS ROSE)

The huge brightly coloured thorns are sure to bring an exclamation from all who see it. These are up to one inch in length, thin, and translucent crimson, and are set thickly along the strong young growth. They are the feature of this species, and well worth growing for the novelty or for floral arrangements. The little cream flowers are botanically interesting as they have four petals instead of the usual five, hence the common name of Maltese Cross Rose.   4b. s. S.

Sericea HEATHER MUIR (variety of *sericea*)

In this recent introduction, the single white blooms are over two inches in diameter and flower over an extended period on a large open bush of fine ferny foliage. This variety sets small red hips.   4c. s. S.

SETIGERA (PRAIRIE ROSE)

With two-inch diameter single blooms of deep rose-pink, ageing paler, and with large soft green blackberry-like foliage, it is noted mainly for its frost hardiness.   5c. s. S.

SETIPODA

This shrub resembles *moyesii* in growth. It bears clusters of pale lilac-pink blooms and long flagon-shaped hips. It was introduced prior to 1895.   3d. s. S.

SOULIEANA

A stiff shrub-climber, noted mainly for its unusual blue-grey foliage and stems. It flowers late in the spring with a concentrated mass of small creamy-white blooms.   5c. s. S.

SPINOSISSIMA ALTAICA

One of the Scotch or Burnet Roses, although this species originated in Siberia. The rather short thicketing bush is arched down by a mass of creamy-white flowers with apricot stamens, set closely along the branches. It flowers early in spring, followed by hips of attractive maroon-black. 2c. s. S.

Spinosissima STANWELL PERPETUAL (*spinosissima* × AUTUMN DAMASK)

A remarkably useful shrub, being truly perpetual, and

almost evergreen. A very airy but neat bush. The flowers, carried close along the stems, are very pale pink to white, attractively crinkled, and double. It was introduced in 1838. 3c. d. P.

Spinosissima WILLIAM III (hybrid)
A quite distinctive plant, being small and twiggy with very small light green leaves and equally tiny blooms of magenta-crimson. It is an excellent foreground species.   1c. sd. S.

STELLATA MIRIFICA (Sacramento Rose)
Another botanically marginal rose, with distinct gooseberry-like grey foliage. The bush is neat, upright, and finely twiggy, with beautiful large cistus-like purple flowers with yellow stamens. It commences flowering late in spring and continues well into summer. The hips are green, rounded, and prickly. It is a valuable collector's piece, but not beyond the average rose grower.   3b. s. S.

WILLMOTTIAE
A huge airy shrub which makes a good background plant, with fine arching branches and light foliage. It flowers early in spring with masses of small lilac blooms set closely along the branches.   5c. s. S.

WOODSII FENDLERI
This rose was first cultivated in 1888, and is another species of the *moyesii* and *cinamomea* group, with single soft-lilac-pink blooms. The relatively tidy thicketing habit and the crop of red flagon hips make it a useful garden shrub.   3b. s. S.

CLIMBERS

The climbing species and similar climbers are grouped here for convenient reference. Compared to modern sported climbers, their habit of growth is usually more satisfying, with thicker lax foliage and more easily controlled growth. Some are extremely vigorous, up to thirty or more feet across, while others are prostrate in habit, scarcely rising from the ground unless supported, but ideal for cascading effects and ground covers.

When pruning, bear in mind the features that you wish to encourage, as outlined for the Species group.

AGLAIA (Yellow Rambler, Multiflora Rambler)
The flowers comprise clusters of small, double rosettes of soft straw-yellow blooms quickly paling to white. The small glossy foliage makes a dense cover on a moderate climber. Climber. d. S.

ANEMONOIDES (Anemone Rose, *sinica anemone*, Pink Macartney)
Actually this climber has no connection to the true Macartney (which is *bracteata*) but is probably a *laevigata* × *odorata*. This flowers even earlier than *laevigata* with large single cerise-pink blooms, much paler on the reverse. The foliage is less dense and luxuriant, and is also less vigorous in growth than *laevigata*.   Cl. s. S.

Anemonoides RAMONA (Red Cherokee)
A sport of *anemonoides* with much deeper colouring, being cerise-red with a paler reverse. This variety also flowers very early in spring.   Cl. s. S.

BANKSIA LUTEA (Yellow Banksia)
Brought from China in 1824, it has been popular for many

years. It flowers very early—from early September to the end
of November—with short clusters of tiny primrose yellow
blooms which festoon the arching canes. If left unchecked it
will develop into an enormous size, but it can be pruned back
as desired after flowering. The thornless canes and foliage are
smooth and light green, and are easily trained into an attractive
dense screen. Being so flexible and thornless, it is ideal for
planting where the comfort of passers-by is to be considered.
It has only a slight scent.   Vigorous Cl. d. S.

BANKSIA BANKSIAE (*alba-plena*, WHITE BANKSIA)

It is similar in most respects to the Yellow Banksia but is
double white with a strong violet scent.   Vig. Cl. d. S.

BLEU MAGENTA

True to its name, the colour of this rose is rich blue-magenta'
with two-inch diameter semi-double blooms in small clusters.
The plants is of shrubby-climbing habit with large glossy
foliage.   5. ss. S.

BRACTEATA (MACARTNEY ROSE)

This vigorous, almost prostrate, climber would make a
wonderful cover for tree stumps, rock outcrops, or creek
banks, for it is both decorative and hardy. It flowers intermittently into autumn with large silky white single blooms
contrasting the orange-yellow stamens. Prostrate Cl. s. R.

Bracteata MERMAID

A hybrid of *bracteata*, and a good climber in any category,
but outstanding as a single rose. The large sulphur-yellow
blooms with amber stamens are borne in tight clusters throughout the season. The foliage is dark glossy green and very
healthy. Mermaid needs plenty of room, as it is very vigorous
and thorny. Introduced in 1917, it has remained as one of the
outstanding roses ever since.   V g. Cl. s. P.

BRUNONII (*moschata nepalensis*, HIMALAYAN MUSK ROSE)

Originally imported into Australia as *moschata*, it is a
rampant climber which bursts into a mass of single small soft
white blooms late in spring. The fragrance is sweetly musk,
while the hips that follow are small, orange-red.   Vig. Cl. s. S.

FÉLICITÉ ET PERPÉTUE (*sempervirens* rambler)

A vigorous but lax and easily trained rambler, it bears small, double creamy-white pompons in clusters. It has a distinct fragrance, and dates from 1827.   Cl. d. S.

FILIPES "KIFTSGATE"

This is a rampant climber which resembles the *moschata* types with huge sprays of small soft white single blooms with yellow stamens, followed by equally large heads of tiny red hips.   Vig. Cl. s. S.

FORTUNEANA (probably *banksia* × *laevigata*)

This climber was introduced in 1850. The loose pompons of purest white are set against a glossy green and almost thornless plant. It is a large cascading climber, at its best running free over sheds or trees rather than being formally trained.   Cl. sd. S.

FORTUNE'S DOUBLE YELLOW (BEAUTY OF GLAZEN-WOOD, GOLD OF OPHIR)

This rose is not accurately named, for it is not truly yellow, but a striking buff-orange, flashed with crimson. It is best left unpruned until it develops into a moderate twiggy climber. Small Cl. sd. S.

GIGANTEA

This species from China had an important influence in the early yellow Noisette and Tea roses, and was later used by Alister Clarke, the well-known Australian breeder of Lorraine Lee, Nancy Hayward, and many others. Almost evergreen and perpetual flowering, the soft buff-yellow single blooms are unfortunately rather fleeting. It sets round green hips Vig. Cl. s. P.

GOLDFINCH

First introduced in 1907, Goldfinch is a cascading shrub with small dense foliage, and bears small soft yellow blooms paling to cream. It flowers late in the spring and is scented. Sm. Cl. ss. S.

HELENAE

The huge loose sprays of single white blooms flower late in the spring on this moderately vigorous climber. The autumn display of small red hips are worthwhile in themselves. Cl. s. S.

LAEVIGATA (Cherokee Rose, *sinica alba*, White Macartney)

Introduced from China in 1759, it flowers very early with the most exquisite pure white single blooms, large and delicately rippled at the edges. They bear a resemblance to *bracteata*, the Macartney Rose, hence the rather misleading synonym in Australia. The climber, which is almost evergreen, forms a dense mass of glossy light green foliage. This species is a truly breathtaking sight with which to herald spring.   Cl. s. S.

LAWRENCE JOHNSTON (hybrid *foetida*)

Intense warm yellow blooms are borne for an extended season on a climber with light glossy green foliage. It has a heavy *foetida* scent, and was introduced in 1923.   Cl. ss. S.

LEONIDA (Marie Leonida) (*bracteata* × *laevigata*)

In cultivation since 1832, and one of the rare hybrids of either parents, it has glossy foliage on slim arching and branching canes. The flowers are large, double, and cupped, of soft cream, and flower over a long period.   Cl. d. R.

LONGICUSPIS

Masses of small single white blooms and small red hips are borne on this vigorous but prostrate, almost evergreen climber. Prost. Cl. s. S.

MAX GRAF (*rugosa* × *wichuraiana*)

This is a very valuable ground cover plant, having the dense green foliage and large single lilac-pink flowers of *rugosa* and the long lax climbing canes of *wichuraiana*. It flowers in the spring and is well covered by dense foliage for the remainder of the season. It was introduced in 1919.   Prost. Cl. s. S.

MOSCHATA (Musk Rose)

Botanically, this has been one of the most important rose parents in rose history. Functionally, it is a very satisfying species, being a modest climber which flowers throughout the season with sprays of small loose single to semi-single creamy-white flowers. It is of course musk scented, and sets tiny orange hips.   5c. sd. P.

MULTIFLORA JAPONICA

A strong rambling shrub, almost thornless, it is useful for

forming large impenetrable hedges. In fact, it has been used as a protective fence on dangerous highway curves. It bears masses of tiny white blooms in spring.   5d. s. S.

MULTIFLORA WATSONIANA
Introduced from Japan in 1870, it is a curious mutation, for the leaves are long and willowy, resembling a Japanese Maple. It is a very appropriate plant for Japanese style gardening. 3d. s. S.

PAULII (*arvensis* × *rugosa*)
One of the most valuable of all ground covers, it will make a dense mass of foliage and flowers up to ten or twelve feet in diameter in the spring. The large single butterfly-like white flowers are airy and delicate, contrasting with the thick large *rugosa*-like foliage. It may be trained upwards to over five feet, or allowed to cascade down vertically. The perfume is similar to cloves.   Prost. cl. s. S.

PAULII ROSEA (*paulii* variant)
Similar to *paulii* in its virtue as a ground cover, this variant is somewhat less vigorous. The flowers are clear rose-pink and scented of cloves.   Prost. cl. s. S.

SEA FOAM
This modern introduction (1964) has been variously classified as Floribunda, Shrub, and Climber. Modest climbing canes of up to six feet are equally at home trailing down as they are when supported upwards, thus making it suited for ground cover use, shrub training, or weeping. It bears masses of small white blooms on a glossy green plant throughout the year. A most versatile variety.   Cl. d. P.

SEVEN SISTERS (*multiflora platyphylla*)
This rose received its name from the loose clusters of blooms which vary between lilac-crimson to lilac-white, literally up to seven different shades. It is a moderate climber, mainly of novelty value, which was introduced in 1817.   Cl. sd. S.

SILVER MOON
A hybrid *laevigata* (1910), with large semi-single blooms of a creamy-white colour, strongly scented, and slightly recurrent. The very vigorous climber has dense glossy foliage. Vig. cl. ss. R.

TAUSENDSCHÖN (THOUSAND BEAUTIES)

A concentrated mass of delicately loose rose-pink blooms, two inches in diameter and cupped, which age to white are featured on this modest climber. It has light green foliage and is almost thornless, and was introduced in 1906.   Cl. sd. R.

TEA RAMBLER

A stout climber (1904) which smothers itself with blossom in the spring. The flowers are medium size rose-pink, paling as they open, with a tea scent. The small bronze foliage makes a dense cover.   Cl. sd. S.

VEILCHENBLAU (VIOLET-BLUE)

Given cool weather and light shade, this rose is the nearest to true blue. The crimson buds open violet blue with a white centre, semi-single, and scented. The bush is thornless and of modest climbing habit. It was introduced in 1909. Sm. cl. ss. S.

WEDDING DAY

A comparatively recent (1950) hybrid climber, renowned for the huge mass of blooms in late spring. The two-inch single blooms of creamy-white have a particularly large centre of orange-yellow stamens. The sprays of blooms and subsequent red hips are large, borne on a strong climber with large glossy foliage.   Vig. cl. s. S.

WICHMOSS

As the name implies, this is a hybrid between *wichuraiana* and the Moss "Salet." The buds are moderately mossed, with small crests on the sepals. The two-inch blooms are semi-double bluish pink, opening to expose golden stamens. Growth is of prostrate or climbing habit, while the fragrance is spicy. Cl. sd. S.

WICHURAIANA (MEMORIAL ROSE)

Known mainly as the parent of a class of climbers, and indirectly, some Floribundas, it nevertheless makes a strong spectacular ground cover. It is completely prostrate, almost evergreen with small rich green foliage, and has an extended display of late spring blooms. These are small, dainty, and white, with strong apple scent.   Prost. cl. s. S.

Wedding Day, a recent hybrid climber

Raubritter, a low, spreading Macrantha

Frau Dagmar Hastrup, a small Rugosa

Geranium, a Moyesii variety

Wichuraiana ALBERTINE
Although not typically *wichuraiana*, this rose is one of the most beautiful, with large coppery-chamois blooms, very double and flat, and moderately stout climbing habit. It is richly perfumed and well remembered as one of the most popular old roses in England (1921).   Cl. d. S.

Wichuraiana ALBERIC BARBIER
This rose possesses the best virtues of the *wichuraiana* hybrids, strong prostrate growth, clean glossy foliage, and sharp apple perfume. The shapely soft yellow buds open to creamy-white double blooms with rich perfume (1900). Vig. cl. d. S.

Wichuraiana AVIATEUR BLÉRIOT
The small but well shaped blooms of soft warm yellow flower singly or in small clusters. Of average climbing vigour, the foliage is dense and glossy (1910).   Cl. sd. S.

Wichuraiana BLOOMFIELD COURAGE
One of the brightest of the old climbers (1925), the small single blooms, crimson with a white eye, flower singly or in small clusters closely set along the stems. Being of moderately vigorous habit, with small foliage and almost thornless stems, it is ideal as a weeping standard or as a cascading climber. Cl. s. S.

Wichuraiana DEBUTANTE
This hybrid is very similar to the better known "Dorothy Perkins," but with smooth dark green foliage that is more resistant to mildew. The small cameo-pink blooms in small clusters are cupped, but finally open to ruffled reflex double blooms that are softly scented (1902).   Cl. d. S.

Wichuraiana DOROTHY PERKINS
This popular rose features small bright rose-pink blooms in clusters on a vigorous and completely prostrate climber. However it is subject to mildew at times (1901).   Cl. sd. S.

MODERN SHRUBS

ALTHOUGH this group brings together the roses that cannot logically find a place in other groups, it is by no means with derogatory intention. Indeed, the opposite is the case, as many varieties here have persisted and become established despite the disadvantage of not belonging to a conventional class. Hybrid Musks form a large part of the group, and also the "Frühlings" group. Several quite popular modern roses such as Iceberg, The Fairy, and Scarlet Queen Elizabeth are included for the simple reason that they are often omitted from popular catalogues because they do not "fit" into conventional selections.

Prune as for modern roses, although lighter, retaining more branches to encourage bushy growth.

AUTUMN DELIGHT (hybrid MUSK)
This rose must surely be a top recommendation as it is scarcely ever without flowers. They are pointed, creamy-apricot in the bud opening to form clusters of large creamy-white semi-single blooms. The bush is of neat modest size and has very few thorns (1933). 3b. ss. P.

BERLIN (hybrid MUSK)
For sheer colour, this is one of the brightest Musks, with large loose single blooms of rich scarlet borne in clusters. It has moderate perfume and stout upright growth (1944). 4b. s. P.

BONN (hybrid MUSK)
A very decorative display of large loose orange-scarlet blooms in clusters is the feature of this shrub rose. It flowers intermittently throughout the season with a light musk scent (1950). 4b. sd. P.

BUFF BEAUTY (hybrid Musk)

One of the most popular "Musks" since its introduction in 1939, and deservingly so. The flowers are neat pompons of apricot-buff and are delightfully scented. The small clusters of blooms nod gracefully from the bush, which is moderately spreading, dense in foliage, and tidy in habit. This is a highly recommended rose.   3c. d. P.

CORNELIA (hybrid Musk)

Sprays of soft apricot-pink, small and rosette shaped, with golden stamens, flower evenly over the bush, which is neat and compact. The blooms are fragrant (1925).   4c. sd. P.

ELMSHORN (hybrid Musk)

Masses of small deep pink pompons with a moderate perfume make a decorative display on a large but tidy shrub (1950).   4c. d. P.

EVA (hybrid Musk)

Mainly of historic interest as the parent of many early Floribundas, the bush is rather too coarse and open for a good garden effect. The three-inch semi-double blooms of cerise-red with pale stamens are borne in large clusters (1933).   4c. sd. R.

FELICIA (hybrid Musk)

Another excellent "Musk" shrub, especially for those who like modern flower forms. The flowers of silvery apricot-pink are of small Hybrid Tea shape, in gracefully arching clusters on a dense bush that seems to cascade outwards with the weight of flowers. The blooms are delightfully fragrant (1928). 3c. sd. P.

FRITZ NOBIS (hybrid *eglanteria*)

This rose is renowned for its concentrated display of bloom late in the spring. The three-inch semi-double blooms with pointed bud formation are soft salmon pink, in clusters that almost cover the stout, mid-green bush. The hips are dull red in colour (1940).   4b. sd. S.

FRÜHLINGSANFANG (Spring Song)

This is another fine single rose, softer in colour than Spring Morn and Spring Gold. The four-inch flowers are ivory-white with apricot stamens. It, too, flowers intermittently on a large dark green bush, setting black hips (1950).   4c. s. R.

FRÜHLINGSGOLD (Spring Gold)

A good display of soft golden blooms open to reveal buff stamens. The bush is hardy, large, and arching (1937). 4c. ss. R.

FRÜHLINGSMORGEN (Spring Morn) (hybrid *spinosissima*)

Spring Morn is a fine example of a single rose. The blooms, large and crisp, are cream, distinctly margined rose-pink, with orange stamens. It flowers intermittently throughout the year from an early start, and sets large maroon hips. The plant is fairly large, but manageable, and is recommended (1942). 4c. s. R.

GOLDEN WINGS (hybrid *spinosissima*)

This beautiful rose features a strong shrub habit and "modern" foliage. The large single yellow flowers remind one of elegant butterfly wings. Being quite perpetual, it is recommended especially for those who like singles (1956).   5b. s. P.

ICEBERG (Schneewittchen)

Although normally listed as a Floribunda, its strong growth and handsome light green foliage make it a must in the shrub group. The pointed buds open to pure white semi-double blooms. This is a magnificent plant by any standard (1958). 5c. sd. P.

MAIGOLD (hybrid *spinosissima*)

This hybrid produces a spectacular display, where the rich green foliage contrasts with the large semi-single blooms of warm gold. The contrasting stamens are amber, and the perfume is spicy (1953).   4c. ss. S.

MARGUERITE HILLING

This is a pink "Nevada" sport, identical in all respects except that the flowers are deep pink. It is equally ideal as a large shrub.   5b. ss. P.

NEVADA (hybrid *moyesii*)

One of the most satisfying large shrub roses, it makes a large but compact bush that is almost thornless, and flowers constantly with very large cream blooms lightly tipped with deep pink. The semi-single blooms have a strong fruity scent (1927). 5b. ss. P.

PENELOPE (hybrid Musk)

A seedling of the well known "Ophelia," the blooms distinctly resemble the parent in a more demure size. The small but well formed semi-double blooms of soft shell-pink are recurrent and fragrant on a thick dark bush (1924). 4c. sd. P.

PROSPERITY (hybrid Musk)

Large clusters of white, rosette-like blooms are featured on this hybrid, introduced in 1924.   5c. ss. R.

RED GLORY

This modern shrub (1958) deserves special mention as a hedging or bedding rose. The three-inch semi-single blooms, light dusky red, flower profusely in small clusters on a bush that is particularly healthy, clean, and compact. It may be pruned by shearing like a hedge.   3c. ss. P.

SCARLET FIRE (Scharlachglut)

Large blooms of particularly rich velvety scarlet contrasted by yellow stamens are borne close along the arching branches. A long spring flowering followed by a crop of red hips makes this a useful acquisition (1952).   5c. s. S.

SCARLET QUEEN ELIZABETH (hybrid Tea)

Too tall to mix with most modern hybrid Teas, it is too good to overlook as a shrub. Large cupped blooms of orange-scarlet are borne throughout the year on a thick plant that is tall and upright (1963).   5b. d. P.

SCHOENER'S NUTKANA (*nutkana* × Paul Neyron)

This rose gives quite a striking display of large four-inch single clear rose-pink blooms on a large arching shrub (1930). 4c. s. S.

SPARRIESHOOP

This 1953 hybrid features large, smooth, rich green foliage and freely bears large single blooms of salmon-pink with yellow stamens, ageing paler.   4b. s. P.

THISBE

Large sprays of two-inch rosette-like flowers of chamois-yellow, and musk scented, are borne on a fairly compact light green bush.   4b. sd. R.

TITIAN

An Australian bred rose (1950) that gives a prolonged display of coral to cerise-pink, with blooms that are quite large and delicately cupped.   4b. sd. P.

VANITY

The huge sprays of rose-pink flowers, almost single, are loosely decorative, although the large open plant appears rather gaunt growing on its own (1920).   4c. ss. R.

WILL SCARLET (hybrid MUSK)

This rose gives a bright display of semi-double light red blooms in loose clusters. It is a large, deep green plant, with a display of orange hips (1948).   4c. sd. R.

RUGOSA ROSES

Species and hybrids of *Rosa rugosa*, the Japanese or Ramanas Rose, are sufficiently numerous and distinct to warrant this separate grouping. Those closely resembling the original species, such as Scabrosa, Blanc Double de Coubert and Roseraie de l'Hay, make wonderful garden plants, with their thick, rough, rich green foliage and dense low habit, their large flowers and large red hips. Others in the group have particular features of exceptional foliage, flowers, hips, or scent.

This group can be used to advantage to blend with modern roses or other shrubs, as individual specimens, or massed in beds or rows. Most are very hardy and disease resistant, especially those closest to the species form. They should be maintained by trimming lightly into shape and removing smothered wood. The more orthodox forms are pruned and maintained in the manner of modern roses.

BLANC DOUBLE DE COUBERT

A very desirable and popular *rugosa*, introduced in 1892, with compact spreading habit and lush green leaves. The flowers are pure white, large, flat, and semi-single, with a distinct fragrance, and are borne steadily throughout the season. This rose is a contrasting companion for "Roseraie de l'Hay." 2c. ss. P.

CALOCARPA (*rugosa* × *chinensis*)

This hybrid is rather more open in growth than a typical *rugosa*, but with moderately large, single, lilac-crimson blooms, and a fine display of medium size bright red hips (1891). 2c. s. P.

CONRAD F. MEYER

More closely resembling a Tea rose, this *rugosa* features

silvery-pink double blooms. The foliage is thick and dark green, but is susceptible to rose rust. The growth habit is rather gaunt.   4b. d. R.

FIMBRIATA (Dianthiflora, Phoebe's Frilled Pink)
This is an interesting variety with frilled, carnation-like blooms of palest pink to white in clusters. It flowers spasmodically in autumn, and sets a few red hips. Although the growth is *rugosa*-like, it is more upright, and is strongly fragrant (1891).   3b. sd. R. .

F. J. GROOTENDORST
The clusters of small rich crimson blooms with finely serrated petals, flower steadily throughout the season. Foliage is leathery, mid-green, with no scent or hips (1918). 3b. sd. P.

FRAU DAGMAR HASTRUP
A small but very impressive *rugosa*, quite dense, low, and spreading in growth. The large single lilac-pink flowers and cream stamens, are followed by large red hips. There is a slight perfume. This rose is a good specimen plant or dense ground cover when closely planted.   1d. s. P.

MICRUGOSA (*rugosa* × *roxburghii*)
This hybrid has very large, single, palest pink blooms in spring, followed by green-orange prickly hips. The plant is thick, angular, and thorny, and makes a good impenetrable hedge. It was introduced prior to 1905.   3c. s. S.

MICRUGOSA ALBA
This is a second generation hybrid, with large white single blooms which repeat later in the season. Growth is more compact than *micrugosa*.   2c. s. R.

MRS ANTHONY WATERER
The loosely double blooms of magenta-crimson are typical *rugosa* colouring. However the bush is open and arching, resembling the Hybrid Perpetual influence. The scent is particularly rich (1898).   4b. sd. R.

NOVA ZEMBLA
This rose is a "Conrad F. Meyer" sport, differing in that its blooms are creamy-white instead of pink (1907).   4b. d. R.

Buff Beauty, a popular hybrid Musk

Centifolia muscosa, or Common Moss

Général Galliéni, a highly recurrent Tea

Foetida persiana, parent of the Pernetianas

PINK GROOTENDORST

This hybrid is a sport of "F. J. Grootendorst," with more striking effect. The small blooms of clear pink are borne in clusters closely resembling carnations or pinks. It flowers throughout the season on an attractive light green bush, making itself equally at home in a species or a formal modern planting (1923).   3c. sd. P.

ROSE À PARFUM DE L'HAY

Large, globular blooms of cherry-red, ageing mauve, with heavy damask scent are a feature of this rose. The blooms are recurrent on a stout, hardy, but not typically *rugosa*-like bush (1903).   2c. d. R.

ROSERAIE DE L'HAY

One of the best examples of a double *rugosa*, this hybrid was first introduced in 1901. The buds are long and elegant, and open through a peony-like stage to blooms four inches in diameter. They are rich crimson-purple and fragrant. There are very few hips. The bush is compact, lush, and dense. 3c. sd. P.

RUGOSA ALBA

This is the single white form of *rugosa*, with large blooms borne in clusters throughout the season, followed by a spectacular display of large scarlet hips. It is one of the outstanding *rugosas* and makes a splendid companion for "Scabrosa." 3c. s. P.

RUGOSA RUBRA

This rose is very similar, if not identical, to *R. typica*, and is the deepest crimson-purple form of *rugosa*. The blooms are large and single, flowering throughout the season and setting a typical fine display of hips.   3c. s. P.

SARAH VAN FLEET

Though of doubtful parentage, this rose appears to be mid-way between *rugosa* and a Hybrid Tea. The blooms are clear pink and intensely fragrant, of loose Hybrid Tea form, on an upright bush of light green glossy foliage.   4b. sd. P.

SCABROSA

A variant of *rugosa*, Scabrosa typifies and emphasises the

best features of the species. The single lavender blooms and bright red hips are large, and the glossy green foliage is lush, making it the leading variety of this group. In fact, it would be the first choice in a species collection, besides being admirably suited to planting as a garden shrub. Massed by itself or with other *rugosas* it will make a low maintenance and impenetrable ground cover or hedge.   3c. s. R.

SCHNEELICHT

This robust spreading and rather vicious bush literally covers itself with medium size single white flowers for several weeks in spring (1894).   2d. s. S.

SCHNEEZWERG (Snow Dwarf)

The outstanding feature of this rose is the anemone-like flowers—two rows of neat, white, symmetrical petals with yellow stamens, which are produced throughout the season. The small bright red hips, dense green foliage, and light scent make it a very useful plant (1912).   2c. ss. P.

SOUV. DE PHILÉMON COCHET

An 1899 sport of the better known "Blanc Double de Coubert," it varies mainly in that the flowers are more double, almost to a crumpled centre, while the centre is softly creamy-pink.   2c. d. R.

CHINA ROSES

In this section are a number of the *R. chinensis* varieties and other close varieties. In addition, a few varieties are included which, although not true Chinas, are more appropriately grouped here than elsewhere. They are all free flowering and fairly orthodox in flowering and growing habit, with rather light twiggy growth and thin smooth foliage. The fragrance is variable.

They are pruned during winter in much the same manner as floribundas and miniatures, thinning out the spent wood and slightly shortening the other branches.

BLOOMFIELD ABUNDANCE

Except for the long lobed sepals, this is almost identical to "Cécile Brunner," and has been sometimes mistaken for such. However, the growth is far stronger; seldom under six feet and up to ten feet if unchecked. Out of a fairly dense, woody shrub, huge airy panicles of bloom burst forth. Altogether, it is a highly recommended shrub (1920). 5b. sd. P.

CÉCILE BRUNNER

The well known and ever popular old "sweetheart" rose (1881), it has perfectly shaped small flowers, only two inches in diameter. They are pink, slightly suffused with yellow, and are borne either singly or in small sprays on a rather woody plant. It will reach about two feet, but an old plant will eventually double that height unless checked. "Bloomfield Abundance" is a similar but taller variety. 1c. sd. P.

CRAMOISI SUPÉRIEUR (Agrippina)

This is a typical example of the China type; loose, cupped blooms on arching stems of a light airy bush with fine bronze foliage. The blooms are a rich crimson in colour with a tea scent, and of course, are perpetually in flower (1832). 1c. sd. P.

ECHO (Baby Tausendschön)

A dwarf sport (1914) of "Tausendschön," this rose has clusters of loose, cupped blooms of rose-pink to white and grows to a modest two to three feet high.   1c. sd. P.

GLOIRE DES ROSOMANES (Ragged Robin)

The large, loose, semi-double flowers of light crimson are borne in large clusters. The plant is rather open but with thick foliage, and was used extensively as a hedge rose. These days it may be used as a good transition into the modern roses (1825).   2c. sd. P.

HERMOSA (Armosa, Melanie Lemaire, Mme Neumann)

Introduced about 1840, this rose is a reversion from the Bourbon class. The globular shape in the lilac-pink blooms, the scent, and the thicker foliage, are all typical of Bourbons. However, the small compact bush, the smaller blooms in clusters, and the freedom of flower are strongly of the China type. In all, a very useful and highly recommended rose, especially for front positions.   2b. d. P.

LITTLE WHITE PET (White Pet)

This rose is possibly a dwarf, recurrent flowering form of the rambler "Félicité et Perpétue" dating from 1879, although it is more at home with the China types. The flowers are compact, white, and rosette like in small clusters, and are borne throughout the year on a low spreading bush of deep green foliage. It is ideal for low edges.   1d. d. P.

MINIMA (Rouletti, Fairy Rose)

The original miniature or fairy rose (1815), and parent of the modern miniatures, it is a worthy variety in its own right. Small, semi-single, rose to carmine-pink blooms, flower throughout the season on a twelve-inch bush with matching tiny foliage. It is sweetly fragrant.   1c. ss. P.

MUTABILIS (Tipo Ideale)

Few species bloom as long and steadily as this, or with such a variance of colour. The large loose single blooms are red in the bud, open to buff-yellow, then through orange to crimson as they fall. The large bush is open and twiggy, producing the blooms on huge panicles.   4c. s. P.

OLD BLUSH (Pink Monthly, Parsons' Pink China)
Introduced into Europe from China about 1760, this and *semperflorens* are the only survivors of the four famous "Stud Chinas" that revolutionised current rose breeding. It commences flowering early and continues throughout the season, bearing medium-sized globular blooms of rose-pink, which pale on opening, then deepen in the sun as the petals reflex outwards. The bush is angular, short, and moderately compact.   1c. sd. P.

PERLE D'OR (Yellow Cécile Brunner)
As the synonym implies, the Perle D'Or resembles a yellow "Cécile Brunner," although the blooms and bush are slightly larger. The colour is apricot to buff, toning well with "Cécile Brunner" and a good companion for it, but equally good in its own right.   2c. sd. P.

SEMPERFLORENS (Slater's Crimson China, and other similar names)
Historically significant as an important parent of our modern roses, this rose features rich crimson blooms which are loose, cupped, and delicately fragrant.   2c. sd. P.

SERRATIPETALA
This rose is similar to *semperflorens*, but has curious serrations on the petals. The blooms are crimson grading to pink at the centre, and fragrant, while the bush is twiggy and spreading.   2d. sd. R.

THE FAIRY (Polyantha class)
Large sprays of small China-pink blooms are borne throughout the year on a low spreading bush of close glossy green foliage. Almost of ground covering habit, it is wonderful for softening garden edges and for low massed effects.   1d. sd. P.

VIRIDIFLORA (Chinese Green Rose)
An extremely interesting and useful curiosity, it has quite normal China rose foliage and habit, with growth to four feet, and it flowers throughout the year. However, in place of petals

viridiflora has sepal-like elements which form two-inch rosette-like "flowers," green on opening, shading bronze as they age. It is very popular with floral artists.   2b. sd. P.

WHITE CÉCILE BRUNNER

A 1909 sport of "Cécile Brunner," the blooms are white to slightly buff-yellow, but otherwise it is the same as the original. 1c. sd. P.

THE "OLD EUROPEAN" ROSES

THE earliest cultivated roses, that is, those of the period up to the early nineteenth century, fall broadly into five classes—the Gallicas, Damasks, Albas, Centifolias, and Mosses—and can conveniently be called the "Old European" types. Although by later standards they appear limited in appeal, nevertheless they awaken a nostalgic feeling of more gracious and formal days.

It is important to remember that many of these varieties were only selected mutations, and not seedlings, as the understanding at that time of seed pollination was very limited, while others were the product of chance self-sown seedlings. As apparent improvements were noticed by the old gardeners, they isolated the new strain and propagated it as another cherished addition to their collection. Later, when the gardeners understood a little more about breeding, many hundreds of scarcely differing seedlings were distributed under a bewildering catalogue of names. Fortunately the inferior varieties were soon discarded, leaving us only with those that survived by popular selection.

There appears to have been only four original species parents involved in developing the "Old European" classes— *gallica*, *damascena*, *phoenicia*, and *moschata*. The perpetual flowering habit that we know today was unknown to these old types, consequently most were only spring flowering, although over an extended period of several weeks. *R. moschata* was the only recurrent flowering parent, giving rise to a few recurrent Autumn Damasks.

Colours of the "Old Europeans" ranged from white, through pinks to plum-red, and maroon, as the yellows, orange, and scarlet-reds were then unknown. The double form of flower was cherished as it is now, and thus predominated. The bushes are thicket-like in habit, from three to seven feet in height, sometimes fairly neat and compact, while others are more lax and lanky. They are hardy against frost, and resistant to Blackspot disease, although some are prone to powdery mildew. They originated in cool climates, but nevertheless seem to thrive in the hotter areas of Australia.

Pruning consists mainly of shortening back the longer canes in winter and thinning the crowded wood. Alternatively, the long canes may be intertwined through their neighbours, or spread against a fence or wall.

GALLICAS (from *R. gallica* [or *R. rubra*], THE RED ROSE, THE FRENCH ROSE, or THE PROVINS ROSE)

Considered the most ancient of the cultivated roses, the Gallicas were grown long before the birth of Christ, and their influence can be traced through most roses up to the present day. They probably originated in the countries to the east and north-east of the Mediterranean, where they crossed, by chance, with the other indigenous species, *R. canina, moschata,* and *phoenicia.*

As a plant they are attractive, with compact, upright, thicketing growth, deep green leaves, and almost thornless. The flowers range in colour from pink to dusky-red to maroon, with several striped variations. When planted as a hedge they may be sheared to good effect instead of pruning.

The original Gallica was most likely *R. gallica officinalis,* while *versicolor* and Tuscany are very ancient. These are still good garden plants, as also are Assemblage des Beautés, Belle de Crécy, Cardinal de Richelieu, Duchesse de Montebello, Président de Sèze, Sissinghurst Castle, Tricolore de Flandre, and Violacae.

ALBA (or WHITE ROSE)

This rose is also very old, dating back to Roman times, where it is recorded as growing in central Europe and the eastern Mediterranean. It probably evolved from the *gallica* × *canina,* or *damascena* × *canina,* and is characterised by strongly scented white to soft pink flowers, light green foliage, and long, smooth, almost thornless canes of upright habit. The plants are usually hardy and disease resistant, growing from four to seven feet tall as a handsome plant.

Pruning consists of centre thinning, while either intertwining the long canes or cutting them to about half their height.

Alba has been traditionally associated with Britain; indeed the ancient name of Albion refers to the indigenous Alba roses, while the historic White Rose of York is, in fact, *R. alba maxima*. The original species are *alba maxima* and *alba semiplena* (which seem to mutate freely one to the other), while Maiden's Blush is very old. Königin von Dänemark, Celeste, and Mme Plantier are some of the outstanding examples of their class.

DAMASKS (*R. damascena*)

Damasks occur in two forms; the Autumn Damask, which is slightly recurrent, and most likely a cross between *gallica* and *moschata*, while the Summer (actually Spring in Australia) Damask would be a cross between *gallica* and *phoenicia*. Damask refers to the heavy damask perfume of the flower. Some claim that the name refers to its origin from the city of Damascus, which is not unlikely, as it originated around the eastern Mediterranean,

The typical bush is more open than the Gallicas, the stems are very thorny, while the foliage is rather grey in colour. Flowers are mainly pinks, with some featuring margined effects.

Some of the original Damasks would be *R. damascena trigintipetala* (Kazanlik), *versicolor* (York and Lancaster), and *bifera* (Quatre Saisons or Autumn Damask). Other good examples of the class are Blush Damask, Celsiana, Leda, Mme Hardy, and Omar Khayyam.

AGATHE INCARNATA (GALLICA)

This rose features a cluster of pale pink blooms of medium size, neatly ruffled into the button-eye, on a bush of matt light green foliage. It was noted prior to 1815.   2b. d. S.

ALBA MAXIMA (WHITE ROSE OF YORK, JACOBITE ROSE, GREAT DOUBLE WHITE)

Although not as perfect as some later Albas, the effect of

the large rippling creamy-white blooms and golden stamens on the tall light green bush is one of considerable garden beauty. It bears oval red hips and is strongly scented.   4b. sd. S.

ASSEMBLAGE DES BEAUTÉS (Gallica)

The bright crimson blooms of this rose form a flat ruffle of petals gathering into a tight button-eye in the centre. It gives an especially bright garden effect with the showy blooms set against the compact dense green bush. It dates from the late eighteenth century.   2b. d. S.

BELLE AMOUR (Alba-Damask hybrid)

Dainty cup-shaped soft coral-pink blooms show prominent yellow stamens on this rose. The fragrance is spicy, and the bush is dense and dark green.   3b. sd. S.

BELLE DE CRÉCY (Gallica)

The large blooms are cerise-pink in colour, ageing through rich violet to lavender-grey, yet showing silver through the button-eye. The blooms are very fragrant and borne on an arching, almost thornless bush.   2c. d. S.

BELLE ISIS (Gallica)

The pale, slightly salmon-pink petals of this rose are held within a circular cupped outline, on a bush that is thick but gently arching (1845).   2c. d. S.

BLUSH DAMASK

A strong dense twiggy bush, quite hardy and better suited for hedges or background planting rather than for individual blooms. It flowers freely with small loose mauve flowers which open to soft lilac-pink, and is strongly scented.   3c. sd. S.

CAMAIEUX (Gallica) 1830

On this aptly named rose, the camellia-like blooms of attractive striped formation are rose-purple, splashed and striped white, with golden stamens, becoming more muted with age. The bush is compact and dense, but small.   1c. sd. S.

CARDINAL DE RICHELIEU (Gallica)

A popular old rose (1840) and a good choice in its colour, which is deepest violet-purple, showing silvery-white at the base and reverse of the petals, which reflex into a neat ball.

The bush is of medium height with large smooth leaves.
3c. sd. S.

CELESTE (Celestial) (Alba)
For sheer delicate beauty, this rose is supreme. It is clear soft pink, and although semi-double, the large blooms open so crisp and airy, eventually revealing golden stamens, that it is truly "out of this world." A tall grey-green bush sets off the blooms. 4b. sd. S.

CELSIANA (Damask)
This rose has large delicately loose blooms of soft pink, revealing yellow stamens. The large lax bush is grey-green in colour. It was first recorded in the early eighteenth century. 3b. sd. S.

CHARLES DE MILLS (Gallica)
The large very double blooms are crimson-purple, and open through a cup-shaped to a large plate formation, while the bush is dense, compact, and dark green. 2c. d. S.

CHLORIS (Rose du Martin) (Alba)
From almost pointed buds the pale pink blooms reflex with the familiar button-eye. The bush is dense and quite thornless. 3b. d. S.

CONSTANCE SPRY (hybrid Gallica)
Although modern (1961) and not a typical Gallica, it has nevertheless captured the old world charm. The blooms of rich pink are large and neatly incurving, eventually showing their golden stamens, making this undoubtedly one of the most beautiful "old" roses. The foliage is thick deep green, while growth is willowy and upright. The blooms are strongly scented. 4b. d. S.

CORALIE (Damask)
Large globular blooms of soft pink are borne on a bush that is unfortunately inclined to be small, open, and thorny. 2c. d. S.

DAMASCENA VERSICOLOR (York and Lancaster)
An ancient rose, it was named in 1551, and is of considerable historic interest as its name commemorates the English War of

the Roses. The rather loose semi-double flowers are either white or pink, sometimes portions of each, while the plant is open, with soft grey foliage and prickly stems. It is very similar to and possibly originates from *trigintipetala*. However, it is not to be confused with the finely striped *gallica versicolor*.   3c. sd. S.

DUC DE GUICHE (GALLICA)
This rose features beautifully formed blooms of light crimson-purple, opening through cupped shaped to reflexed, with a small green eye.   2b. d. S.

DUCHESSE D'ANGOULÊME (GALLICA, probably *gallica × centifolia*)
The rather loose globular nodding blooms are of blush pink, with large tissue-petals and a strong perfume (1836). 2b. sd. S.

DUCHESSE DE BUCCLEUGH (GALLICA)
This 1846 rose features large cupped blooms opening to flat, of a rich magenta-pink colour, paler to the edges and reverse.   2b. d. S.

DUCHESSE DE MONTEBELLO (GALLICA)
The beautifully formed and refreshingly clear pink blooms of this rose are cupped and with button-eye, and are borne on long waving branches. Altogether it is a fine example of its class.   3b. d. S.

FÉLICITÉ PARMENTIER (ALBA)
The large flat blooms of this rose are palest flesh pink, strongly scented, and are borne on a bushy grey-green bush (1836).   2c. d. S.

FRANCOFURTANA (EMPRESS JOSEPHINE, FRANKFORT ROSE)
This *gallica × cinnamomea* hybrid, which was recorded before 1600, more closely resembles the former. It is not an ideal garden rose. The small rather bushy plant bears loose semi-double blooms of mauve-pink, veined deeper. 2c. sd. S.

GALLICA VERSICOLOR (ROSA MUNDI)
A very old but well known sport of "Officinalis," this rose was described as early as 1581. The habit of growth and flower is the same as "Officinalis." However the blooms are striped

light crimson and pale pink, with prominent yellow stamens. It makes a good low garden display. It should not be confused with the taller "York and Lancaster."   1c. sd. S.

GEORGES VIBERT (Gallica)
This is a rose with a subdued striped effect, being light crimson on light pink. The late flowering blooms are camellia-like, with strong scent, and dense dark green foliage (1853). 3b. sd. S.

HEBE'S LIP (Rubrotincta, Reine Blanche)
Hebe's Lip is a damask, possibly an *eglanteria* hybrid. The charmingly simple cupped semi-single blooms are creamy-white, finely edged with crimson.   2b. ss. S.

HIPPOLYTE
Apparently a hybrid Gallica, this plant is quite tall and thorn-less, with small but dense leaves. The blooms are small reflexing balls of cerise to violet, sometimes grey, and well scented. 3b. d. S.

ISPAHAN (Pompon des Princes, Rose d'Isfahan)
This rose features clusters of clear pink blooms, attractive from the tight bud to the neatly reflexing pompons and distinct button-eye. It flowers freely over a long period. It is a Damask, noted prior to 1832.   2b. sd. S.

JENNY DUVAL (Gallica)
The colour of this rose is its oustanding feature. A striking combination of grey-lavender tones, being deep lilac-pink, shading to white on the reverse, but opening through tones of cerise to grey-brown. The blooms are ruffled, informal, and double.   2b. d. S.

KÖNIGIN VON DÄNEMARK (Alba)
This rose has very double blooms of delicate salmon-pink, cupped when opening but finally reflexing to traditional quartered formation. The rather open plant has matt blue-green foliage (1826).   3b. d. S.

LA VILLE DE BRUXELLES (Damask)
The large blooms of clear pink are tinted with violet and ruffled into a button-eye (1849).   3b. d. S.

'LEDA (Painted Damask)

The blunt, almost deformed buds of brick-brown open to soft white with a characteristic narrow margin of crimson to each petal. The double reflexing and button-eye blooms weigh down the arching stems of the dark green plant. Although variable, it is spectacular at its best.   2c. d. S.

MAIDEN'S BLUSH (GREAT) (Gallica)

An ancient rose, growing wild in Europe, and in cultivation since the fifteenth century, it is a strong hardy free-flowering plant with informal flowers of soft blush pink and sweetly scented.   4c. sd. S.

MARIE LOUISE (Damask)

The large cerise-pink blooms, which age paler, arch down the lax canes of dark green foliage.   2c. d. S.

MME HARDY (Damask)

Surely this is one of the most beautiful of the old-world roses. The large white flowers open through cupped to flat formation within a circular outline, and consistently show distinctly quartered petals and a green eye. The matt grey-green foliage suggests some *centifolia* breeding (1832). 3c. d. S.

MME LEGRAS DE ST GERMAIN (Alba)

The large flat camellia-like blooms of this rose are ivory-white, opening to cream in the centre. The bush is almost thornless, with an arching growth of soft matt green (1846). 3c. d. S.

MME PLANTIER (possibly *alba* × *moschata*)

While this rose is not a true Alba, the large sprays of soft white flowers are sheer perfection. The blooms are rather small, flat, and tightly quartered, rivalling the larger "Mme Hardy" in this colour. The flowers contrast against the dense soft grey-green foliage, producing a delightful garden effect (1835). 3b. d. S.

MME ZÖETMANS (Damask)

The small cup-shaped buds of this rose open to reflexed balls of softest flesh pink, neatly ruffled, with tiny green centres (1830).   2c. d. S.

OFFICINALIS (APOTHECARY ROSE, RED ROSE OF LANCASTER,
RED DAMASK [a misleading title], *splendens*)

An ancient Gallica, cultivated perhaps as early as 3,000 B.C.,
this rose nevertheless makes a worthy garden plant. The semi-
double light red blooms show prominent yellow stamens, and
are held erect from the moderate thicketing bush. The round
hips are brick-red.   2c. sd. S.

OMAR KHAYYAM (DAMASK)

Selected in 1893 from seeds taken from a rose growing on
the grave of Omar Khayyam in Nashipur, this rose is pale pink,
with small star-like button-eyed blooms in clusters. The
thorny upright plant is rather sparse and is light green. 3b. sd. S.

PETITE LISETTE (DAMASK)

This rose, recorded in 1817, features beautiful small rosette-
like blooms of pale pink softly striped deeper, with distinct
button-eyes, on a small bush of light matt green.   1c. d. S.

POMPON BLANC PARFAIT (ALBA)

The blooms of this rose are seen as large clusters of small
pompons, white to pale pink. The plant is dense and upright,
and the foliage is pale green (1876).   3b. d. S.

PRÉSIDENT DE SÈZE (GALLICA)

A quite striking effect in the "old world" tones are featured
in this rose, being magenta in the centre of the large muddled
and cupped blooms, shading to soft lilac at the edges (prior
to 1836).   2b. d. S.

QUATRE SAISONS (AUTUMN DAMASK, *damascena semperflorens, bifera*)

This is one of the earliest cultivated roses, and was extensively
grown in Roman times. Although not a heavy flowerer by
later standards, the spasmodic autumn blooms were nevertheless significant in their time. The loose semi-double crumpled
blooms are borne in close clusters on a bush that is light green
and upright, though thorny. The blooms are well scented.
3b. sd. R.

SISSINGHURST CASTLE (GALLICA)

An apparently unknown old rose that was rediscovered in
the grounds of the castle in 1947, and so named. A neat low

dark green bush which freely bears blooms of purple to deep plum with distinct golden stamens. It is a good foreground plant.   2c. d. S.

SURPASSE TOUT (GALLICA)

The large blooms are of a particularly rich crimson, ageing to cerise pink, and lightly reflexing with a large button-eye. It was first recorded prior to 1832.   2b. d. S.

TRIGINTIPETALA (KAZANLIK, *damascena trigintipetala*)

This rose is best known as the main source of rose attar, and is grown extensively for this purpose at Kazanlik in Bulgaria. The clear pink blooms are loose and semi-double compared to later Damasks, and are very free flowering, while the bush is large and loosely spreading.   4c. sd. S.

TRICOLORE DE FLANDRE (GALLICA)

This aptly named rose has an interesting striped combination with the double reflexing blooms finely striped with blush, lilac-pink, and crimson-purple. It is rather stronger in growth than "Camaieux," but not such a striking contrast (1846). 2c. d. S.

TUSCANY (OLD VELVET) (GALLICA)

One of the best dark colour roses, being rich velvety maroon, contrasted with yellow stamens on the flat semi-double flowers. The growth is upright and thicketing with dense, dark green foliage. Tuscany Superb is rather more double, yet still showing some of the distinctive stamens.   2b. sd. S.

VIOLACAE (MAHEKA) (GALLICA)

A rose with simple charm in the almost single blooms of crimson-purple with golden stamens, flowering heavily for a short period on a slim upright bush.   3a. ss. S.

Mme Hardy, an old-world Damask

Paul Neyron, a Hybrid Perpetual

Quatre Saisons, an ancient Autumn Damask

From left: Belle de Crécy, a Gallica; James
Mitchell, a Moss; and Assemblage des Beautés, a Gallica

CENTIFOLIAS AND MOSSES

CENTIFOLIAS (*R. centifolia*, CABBAGE ROSE, PROVENCE ROSE [not PROVINS ROSE])

This group is one of the most confusing of the old roses. They appear to be of relatively recent origin, having been developed in Holland during the seventeenth and eighteenth centuries from four original parent species, *gallica*, *moschata*, *damascena*, and *alba*. They are not an ancient rose of Roman and Greek times as is sometimes suggested. The name "centifolia" literally means "rose of a hundred leaves," and describes the very double nature of the flowers. This is brought about by the chance mutation of many of the stamens into petals. The subsequent reduction in the number of stamens explains why Centifolias are almost sterile. Many people use the term "cabbage rose" for any large, globular, very double blooms, but it is more likely than not that they refer to some Bourbons or Hybrid Perpetuals, especially the huge pink "Paul Neyron."

It is difficult to generalise a description, but they are often rather lanky, and arching, with the large, heavy blooms weighing the stems over — a rather attractive effect.

Pruning simply consists of shortening the long canes by a third or half.

Later breeding has scarcely improved the original species, so that Centifolia and its variants are still recommended. These are *R. centifolia, bullata, centifolia cristata* (Crested Moss or Chapeau de Napoléon), while others worthy of trial would be Fantin-Latour, Rose de Meaux, Tour de Malakoff, and Robert le Diable, which is slightly recurrent.

MOSSES (*R. centifolia muscosa*)

These roses get their name from a curious effect caused by a mutation which enlarges the glandular projections on the sepals and stalks, giving the characteristic moss-like effect. Centifolias have been the parents of most of the Mosses, although some are from the Damasks. Several modern Mosses have been hybridized but these lack the traditional charm. As they originate by mutation, it is not uncommon for Mosses to mutate back again, or even have both effects on one bush.

The elongation, winging, and bearding of the sepals of roses has fascinated gardeners through the ages.

Study the five sepals, how they unfurl from the bud, and then consider the ancient rhyme,

> Five brothers take their stand
> Under the same command.
> Two darkly bearded frown,
> Two without beards are known,
> While one sustains with equal pryde
> His sad appendage on one side.

Is this not particularly true of the moss roses?

Pruning, as with the Centifolias, is light, merely shortening and tidying the plant.

Likewise with Centifolias, the original species are among the best varieties — *R. centifolia muscosa* (Common Moss), *centifolia muscosa alba* (Shailer's White Moss), and Japonica. However do not overlook A Longs Pédoncules, Comtesse de Murinais, Nuits de Young, and William Lobb.

A LONGS PÉDONCULES (Moss)
This beautiful rose has small loose but well shaped soft lilac-pink blooms, borne in clusters on long arching stems (1854).   3c. sd. S.

BARON DE WASSNAER (Moss)
This rose produces clusters of light crimson blooms ageing rich purple and paling to silver at the centre and reverse (1854).   3b. sd. S.

BULLATA (Lettuce-leaf Rose)
A variety of *centifolia*, the Lettuce-leaf Rose features larger wrinkled (bullate) leaves which are toned bronze, especially in autumn.   2c. d. S.

CENTIFOLIA (CABBAGE ROSE, PROVENCE ROSE)

The original of the class, it was first recorded in cultivation about 1600. The flowers, up to three inches in diameter, are double with the circle of reflexing outer petals containing the ruffle of inner petals, so that the stamens are seldom seen. The blooms are clear rose-pink in colour, and moderately fragrant. The plant is rather lank and open with coarse grey-green foliage.   2c. d. S.

CENTIFOLIA MUSCOSA (COMMON MOSS)

The first mutation from the original *centifolia* was recorded prior to 1750. The double quartered pink blooms are the same, but the moss-like projections cover the sepals, calyx, and stalks, and are fragrant when bruised. This would be first choice of the Moss roses.   2c. d. S.

CENTIFOLIA MUSCOSA ALBA (SHAILER'S WHITE, WHITE BATH, CLIFTON MOSS)

A white mutation of the Common Moss rose, with creamy-white to pure white blooms of similar form and habit to the original. It is well recommended.   2c. d. S.

CENTIFOLIA VARIEGATA

As the name implies, this rose is a striped sport with white flowers splashed and striped with pale pink, although not in striking contrast (1845).   2c. d. S.

CHAPEAU DE NAPOLÉON (NAPOLEON'S HAT, *centifolia cristata*, CRESTED MOSS, CRESTED PROVENCE ROSE)

Another curious and quite famous variation of *centifolia*, where the sepals that cover the bud project large "crests" or "beards" in a fascinating and not unpleasant manner. In fact the name refers to its similarity to the famous cockaded hat. Again, the blooms and habit of growth resemble *centifolia*. 2c. d. S.

COMTESSE DE MURINAIS (DAMASK-MOSS)

This hybrid produces large white blooms, reflexed and with a Damask-like button-eye when open. Heavy pale green moss and winged sepals are featured. It is a rose well recommended (1843).   2b. d. S.

DEUIL DE PAUL FONTAINE (Damask-Moss)
This rose is a rather different type of Moss, with only limited appeal. The globular flowers are purple to maroon, even to mahogany, with stiff moss and very prickly stems. It is slightly recurrent (1873).   2c. d. R.

DUCHESSE DE VERNEUIL (Moss)
The small but rich deep pink blooms on this rose pale slightly to the edges, and are moderately mossed (1856).   2b. sd. S.

FANTIN-LATOUR (Centifolia)
This is one of the best Centifolias, being soft pink with a large display of double reflexing blooms with distinct button-eye, on a bush that is more compact and tidy than most Centifolias.   3c. d. S.

FÉLICITÉ BOHAIN (Moss)
The blooms of this rose are flesh pink, deepening to rose at the centre, double, quartered, and button-eyed. It was first recorded prior to 1866.   2b. d. S.

GENERAL KLÉBER (Moss)
General Kléber features large crisp blooms, clear soft pink in colour and heavily mossed (1856).   2b. d. S.

GOLDEN MOSS (Moss)
This is a relatively modern Moss (1932) with medium sized semi-double blooms of apricot-yellow and slightly recurrent. Although not highly rated, it grows strongly with distinctly aromatic moss on the young branches.   4b. sd. R.

HENRI MARTIN (Red Moss)
By old world standards, this rose has particularly bright crimson blooms, small and semi-double, neatly borne in clusters (1863).   2c. sd. S.

JAMES MITCHELL (Moss)
This rose displays clusters of small double blooms, lilac-pink and well mossed, on a thick deep green bush which makes a good garden display in its class (1861).   2c. d. S.

JAPONICA (Mousseux du Japon) (Moss)
Japonica features the heaviest display of moss of all, not only on the calyx but along the stems and on the leaves, while

the young foliage is attractively tinged purple. The blooms are rather nondescript, magenta-pink, and scented.   2c. sd. S.

LA NOBLESSE (Centifolia)
This rose has clusters of large loose soft pink blooms opening cupped then reflexing, with open sprawling growth (1856). 2c. d. S.

MARÉCHAL DAVOUST (Moss)
The large, flat, well-shaped flowers of this rose are a rich magenta-pink, well mossed, on a bush of rather smooth light green foliage (1853).   2c. d. S.

MME LOUIS LEVÊQUE (Moss)
The blooms of this rose are soft pink, large and ruffled, within a tight cup-shaped formation. The bush is stout and upright with rather "modern" foliage (1893).   2b. d. S.

NUITS DE YOUNG (Old Black) (Moss)
The velvety maroon blooms of Nuits de Young are among the darkest of the old roses. They are small and flat exposing contrasting yellow stamens (1851).   2b. d. S.

PETITE DE HOLLANDE (Centifolia)
True to its name, this rose features small neat flat blooms of rose-pink on a correspondingly small plant.   2c. d. S.

QUATRE SAISONS BLANC MOUSSEUX (*damascena bifera alba muscosa*, Perpetual White Damask Moss)
As the name amply describes, it is a double mutation from the "Quatre Saisons," having changed from pink to white in addition to the mossing, making it a comparatively rare mutation from the Damasks. The flowers are rather loose and informal, but the heavily mossed clusters of buds in spring, and occasionally later, offset this disadvantage. It is heavily scented (1835).   2b. sd. R.

REINE DES CENTFEUILLES (Centifolia)
The blooms produced by this rose are deep pink opening to reflexed, light pink in colour, button-eyed, and ruffled (1824). 2b. d. S.

RENÉ D'ANJOU (Moss)
A large open bush with light foliage, René d'Anjou bears large cupped blooms of soft lilac-pink (1853).   3c. sd. S.

ROBERT LE DIABLE (Centifolia)
This rose features unusual shades of violet-pink blooms,
ageing to purple and cerise—medium sized, and reflexing,
borne on a sprawling bush.   2d. d. S.

ROSE DE MEAUX (*centifolia pomponia*, Pompon Rose)
The dainty cup-shaped blooms open to pompons of soft
pink, on a plant that is small and twiggy, yet upright and neat.
1b. d. S.

THE BISHOP (Centifolia)
This rose features a striking display of medium size rosette-
like blooms of magenta, opening to deep purple, paler to the
reverse and centre.   2c. d. S.

TOUR DE MALAKOFF (Centifolia)
The silvery lilac-pink buds of this rose open to deep tones of
mauve to magenta, with large cupped blooms on a large open
bush (1856).   3c. d. S.

WILLIAM LOBB (Duchesse d'Istrie, Old Velvet) (Moss)
Clusters of crimson-purple to lavender-grey blooms with
well mossed buds are borne on long stout stems. William Lobb
gives a good display on a thick compact bush (1855). 3b. d. S.

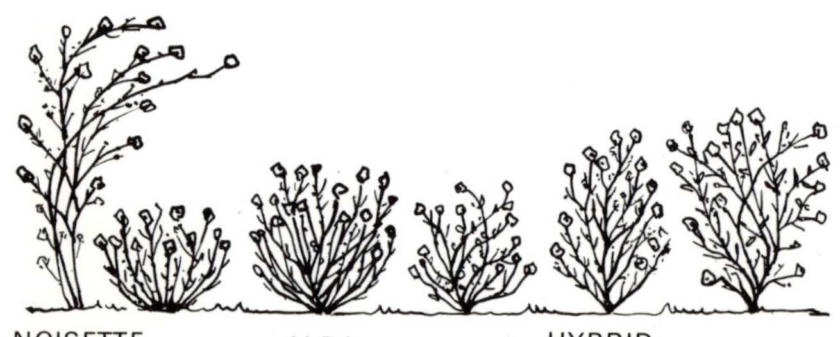

THE INFLUENCE OF THE "CHINAS"

Looking back from our enlightened age, it is incredible to think that such a revolution in modern roses could have been brought about by such a microscopically small factor. I refer, of course, to one particular gene in the chain of characteristics of *R. chinensis*, an insignificant little species from south-east China, the characteristics of which are seen in each of four important roses brought to Europe about 1800.

These four varieties, Slater's Crimson, Parsons' Pink, Hume's Blush Tea-scented China, and Park's Yellow Tea-scented China had several very desirable genes which they were to eventually pass on to the popular roses of the 1800s. These genes produced a new intense red, new yellow shades, a new tea-scent, elegant pointed buds, and light airy habit of growth, but more important than all these factors was the ability to literally flower throughout the year. No sooner did one stem finish flowering than another stem was bursting into new blooms. It is safe to say that every worthwhile modern rose has this vital gene in its makeup, and it has done more than any other factor to make the rose the universal flower that it is today.

The four "Stud Chinas," as they have been called, were natural hybrids of two species; the small *R. chinensis semperflorens*, and a vigorous evergreen from China, *R. gigantea*. Only Slater's Crimson and Parsons' Pink exist today, although each of the four were used extensively in those exciting years in the first half of the nineteenth century.

NOISETTES
The first of the important crosses was between Parsons' Pink China and White Musk, raised by Champney in the U.S.A. It quickly became a favourite shrub in America, posses-

sing the best features from each parent. A nurseryman named Noisette raised some second generation seeds, which he sent to his brother in Paris. They were crossed to other likely species, in particular, *R. multiflora* and *R. chinensis*, Aimée Vibert is well known from these breedings, besides other blush pinks.

Yellow was introduced into the class when the Blush Noisette was crossed with Park's Yellow Tea-scented. From here a quickening succession of yellow Noisettes and yellow Tea roses were raised, prominent among them being Jaune Desprez, Céline Forestier, Lamarque, and Maréchal Niel.

Apart from their historic significance, Noisettes are still unsurpassed in soft yellow and blush tones for perpetual flowering. Bushes are usually of shrub to low climbing habit, and are quite hardy under Australian conditions.

BOURBONS

On the French island of Bourbon in the Indian Ocean, an interesting new seedling was discovered in 1817 growing among the hedge plants of Parsons' Pink China and the Autumn Damask (*bifera*). Known locally as Rose Edouard, it was obviously a chance cross between these two hedge roses. Sensing its potential value, an alert botanist sent it to France where its second generation seedling gave what is regarded as the original Bourbon, "Rosier de l'Ile Bourbon."

Characteristically, Bourbons are strong, stout upright growers, with large cupped, semi-double blooms, and delicious Damask perfume. They flower recurrently from late spring to autumn in colours that range from blush pink to rose-red and later varieties to crimson, including several striped combinations. The class held sway for half a century, finally giving way to the newer Tea roses and Hybrid Perpetuals. Bourbons are the most satisfying of the older types where one wants the charm of the old rose form combined with recurrent flowering.

Rose Edouard and Rosier de l'Ile Bourbon are still available, but are grown mainly for historic reasons. They are excelled by the later La Reine Victoria, Mme Isaac Pereire, Souvenir de la Malmaison, Variegata di Bologna, and Boule de Neige, all of which will make a valuable contribution to any garden.

Most Bourbons grow as a rather upright, compact bush to six or seven feet high, with long relatively thornless canes that may, if desired, be trained espalier fashion against a trellis.

An attractive composition of Old World types

A display of species and their close hybrids

From left: Buff Beauty, Elmshorn, Felicia, and Eva.
Mermaid forms the top background

La Reine Victoria, a typically old-fashioned Bourbon

Pruning of Bourbons approaches that of "moderns"—spuring the flowering laterals after flowering, and shortening the longer, climber-like canes to about half length.

AIMÉE VIBERT (Bouquet de la Mariée) (Noisette)
The sprays of white blossom contrast effectively against the thick glossy green foliage of this moderate climber. The flowers are neat, ruffled, and double, with some scent. Although rather late to start flowering, it is quite recurrent (1828). Cl. d. R.

ALISTER STELLA GRAY (Golden Rambler) (Noisette)
A delightful moderate climber that could well find a place with the moderns, with its shapely little blooms of warm yellow to cream, borne in large open clusters. It flowers consistently throughout the year and is fragrant (1894). Cl. sd. P.

AUTUMNALIS
This rose is of obscure origin, resembling both Noisette and *moschata*. It is nevertheless an outstanding shrub; large, very dense and compact, with clean, matt foliage, and a constant display of loose, creamy-white blooms in clusters.   5c. sd. P.

BOULE DE NEIGE (Bourbon)
Although not typically Bourbon, this rose is still a wonderful shrub which is scarcely ever without flowers. As the name implies, the flowers are pompons of crisp white in small clusters and are well scented. The plant is particularly handsome, being dense and rich green (1867).   4b. d. P.

BOURBON QUEEN (Reine de Iles Bourbon)
The Bourbon Queen produces a decorative display of loose soft magenta-pink blooms subtly striped and flecked paler, on a vigorous semi-climbing bush. It is strongly scented but only slightly recurrent (1835).   4b. sd. R.

CÉLINE FORESTIER (Noisette)
Large creamy-yellow blooms, fully double and quartered, with delicious Tea-scent, are featured on this modest climber. Sm. Cl. d. P.

CHAMPNEY'S PINK CLUSTER (*noisettiana*)
This rose is the original Noisette (1811), being *chinensis* ×

moschata. The clusters of small, cupped semi-double blooms are soft blush, tinged with lilac, and fragrant. The plant is an arching shrub-climber.   5c. sd. R.

CLOTH OF GOLD (CHROMATELLA) (NOISETTE)
A well known climber since 1843, this rose has large sulphur-yellow blooms, pointed in the bud and cupped Tea-rose shape when open, on a moderate climbing plant. It has a rich tea scent.   Cl. d. R.

COMMANDANT BEAUREPAIRE (BOURBON)
The striped blooms of pink, finely streaked with red to purple, are double and cupped, with good perfume (1874). 4b. d. R.

HONORINE DE BRABANT (BOURBON)
This rose produces typically cup-shaped blooms of white to soft lilac which are striped with mauve and crimson. 4b. d. R.

JAUNE DESPREZ (DESPREZ À FLEURS JAUNES) (NOISETTE)
A very early Noisette (1830), yet still one of the best, with especially rich perfume. The moderately large flowers are soft apricot-pink, flat, and double, borne in small clusters throughout the season on a thick, medium sized climber.   Cl. d. P.

LAMARQUE (NOISETTE)
Lamarque is another early Noisette, and has sprays of shapely buds opening to large nodding blooms of creamy-white, with yellow centres, on a modest bush of clear light green foliage. It is very free-flowering and fragrant, and is highly recommended (1830).   5c. d. P.

LA REINE VICTORIA (BOURBON)
If ever there was a rose that to most people was typically "old-fashioned," this is the one. Of distinct globular shape, the outer petals form an inflexing circle, allowing only a small double centre to show when fully open. It is clear pink and richly fragrant, on a tall, semi-climbing bush (1872).   4b. d. R.

LOUISE ODIER (BOURBON)
This rose has large cupped blooms of rich pink, ageing lilac. It is fragrant (1851).   4b. d. R.

MARÉCHAL NIEL (Noisette)

The flowers of this rose are a soft yellow, with the influence of Tea-rose in the fragrance and the large nodding, double blooms. It is a moderately vigorous climber with glossy but sparse foliage (1864). Similar, but preferable to "Cloth of Gold."   Cl. d. P.

MME ALFRED CARRIÈRE (Noisette)

A comparatively late variety of its class (1879), with well shaped creamy-blush buds and large ruffled white blooms borne on a shrub-climber.   Cl. d. R.

MME ISAAC PEREIRE (Bourbon)

This is another fine example of "old fashioned" charm with very large cupped blooms of deep carmine-pink, forming within its circular outline a quartered and ruffled centre. It is strongly fragrant, on a tall arching bush (1881).   4c. d. R.

MME ERNST CALVAT (Bourbon)

An 1888 sport of Mme Isaac Pereire, differing in colour. which is clear pale pink with a somewhat deeper reverse. 4c. d. R.

MME PIERRE OGER (Bourbon)

A sport of "La Reine Victoria," with the same beautifully globular blooms but differing in colour, which is soft ivory in the bud but deepening to lilac-rose as the petals expose to the sun—a delightfully subtle effect (1878).   4b. d. R.

MME SANCY DE PARABÈRE

This is the best example of a Boursalt, a class of climbers which originated about 1820 from a *chinensis* crossed to an unknown thornless species. The Boursalt understock is no relation, being in fact *indica major*. "Mme Sancy" flowers early with a mass of light violet-rose blooms which are borne on long thornless stems (1875).   4b. sd. S.

PORTLAND ROSE (Duchess of Portland)

Possibly Autumn Damask × Slater's Crimson, the Portland Rose has a characteristic thick compact bush with the small clusters of flowers closely nestled back into the pale grey-green foliage. The flowers are semi-single light red, three to four inches in diameter and lightly scented.   2c. ss. R.

ROSE EDOUARD (AUTUMN DAMASK × PARSONS' PINK CHINA) (BOURBON)

Although it is not a good garden plant, it marks the break-through into the Bourbon class, and is of considerable historical interest. The growth is China-like but rather taller; the soft lavender blooms, tinged deeper, are small and globular, but need good weather to open consistently. It is recurrent and fragrant, and was recorded prior to 1819.   3b. d. R.

SOUVENIR DE LA MALMAISON (BOURBON)

This rose is among the most exquisite for shape, colour, and perfume. The flowers are large and double, opening through cup-shaped to flat quartered flowers of soft creamy-blush with rich perfume (1843).

The climbing form is the best known, being quite moderate and suited to pillar training. The bush form reaches about two to three feet high.   Cl. and 2c. d. R.

SOUVENIR DE ST ANNE'S (BOURBON)

A sport of "Malmaison," this rose is semi-single, with gracefully arranged petals of softest blush-pink.   4c. ss. R.

VARIEGATA DI BOLOGNA (BOURBON)

The most striking and possibly the best striped Bourbon, with large globular blooms of crimson-purple striped on white. It is strongly fragrant on a strong arching plant (1909). 4b. d. R.

ZÉPHIRINE DROUHIN (BOURBON)

This rose is not typically Bourbon, as the rich cerise-pink flowers are smaller and semi-double, while the climbing growth has a coppery-purple tone when young. However it provides a striking display as a climber or large shrub, especially where its thornless habit can be used to advantage (1868). Cl. sd. R.

ZIGEUNERKNABE (GIPSY BOY) (BOURBON)

This rose is again not typical of the class, but makes a good display with clusters of small crimson flowers which age purple, on a strong open plant.   4c. sd. S.

TEA ROSES, HYBRID PERPETUALS, AND HYBRID TEAS

THE classification of roses into convenient compartments has seldom been easy, but with the continued influence of the China "blood," this became more complex in the Hybrid Perpetuals (H.P.) and the later Hybrid Teas (H.T.).

The Tea roses originated from various crosses between Hume's Blush China and Bourbons, or Park's Yellow Tea-scented China and Noisettes, giving very recurrent roses in soft shades of cream, yellow, and pink, sometimes frost tender and delicate in growth, but with shapely elegance of form and good perfume.

Hybrid Perpetuals generally originated from Bourbons or Noisettes crossed with various Hybrid Chinas. They were the result of the search for even more recurrent flowering, and reached their heyday during the mid-nineteenth century. In the shape of their flowers they moved significantly towards the "modern" form, with larger inner petals starting to raise the shape of their flowers they moved significantly towards the climbing tendency, also approached the more modern form. Fragrance was a strong feature. However, despite their pretentious title, they were not truly perpetual. This is noticeable by observing the improvement when they were finally crossed with the Tea roses, giving obviously enough, the Hybrid Tea.

La France, traditionally regarded as the first Hybrid Tea, was raised in 1867 with the likely parents being Mme Victor Verdier, a Hybrid Perpetual, and Mme Bravy, a Tea rose. The term Hybrid Tea has remained now for over a century despite the introduction of other breeding strains and the present high activity in modern hybridizing. Hybrid Teas, when crossed with the Austrian Copper were known for a time as "Pernetianas," but these have gradually been absorbed back into the H.T. class. Pernetianas contributed to our present colour range the bright yellows, orange, and bronze.

These later classes perfected blooms with larger reflexing petals, long pointed buds, and a wider and brighter range of colours, together with the quicker repetition of flowers that we know so well today. .

BARON GIROD DE L'AIN (H.P.)

This rose produces a rather striking effect where the cupped wavy petals of crimson-maroon are distinctly edged with white. Although not as contrasting as the better known "Roger Lambelin," it grows and flowers better, and is fragrant, although both are prone to rust disease (1897).   3c. sd. R.

BARONNE PRÉVOST (H.P.)

The particularly large flat quilted blooms over four inches in diameter, are mid-pink ageing to soft lilac. The bush is stout and upright, carrying the fragrant blooms on stout stems (1842).   3b. d. R.

DEVONIENSIS (MAGNOLIA ROSE) (TEA)

This Tea rose has large, double, beautifully formed blooms of creamy-white, tinged apricot to the centre, open flat and ruffled, fragrant and recurrent; altogether a very desirable variety (1838). The climbing form (1858) is a stout open vigorous plant that is better known than the dwarf.   Vig. Cl. or 3c. d. R.

FERDINAND PICHARD (H.P.)

The blooms of this rose are softly striped crimson-purple on a white base, ageing soft purple, which gives a pleasantly mild striped effect. Typically stout upright H.P. growth, and very fragrant (1921).   3b. d. R.

FRAU KARL DRUSCHKI (SNOW QUEEN) (H.P.)

One of the whitest roses of all, although sometimes tinged pink, the blooms are large double and shapely, borne on a tall sprawling light green bush (1901).   4c. d. R.

GÉNÉRAL GALLIÉNI (TEA)

This rose is scarcely ever without flowers. The blooms are large and loose, the petals becoming small and double to the centre, buff-cream shaded coppery-red to crimson, with good tea scent. The bush is of fine twiggy growth but with dense glossy foliage, making a good garden display (1899). 3d. sd. P.

GÉNÉRAL JACQUEMINOT (H.P.)

The fragrant blooms are red, ageing magenta, on long upright stems. It is best known as the parent of many later good reds (1853).   3b. d. R.

GEORG ARENDS (H.P.)

The large shapely pointed blooms are of soft clear pink, strongly fragrant, and are borne on stout almost thornless canes with rich green foliage (1910).   3b. d. P.

GLOIRE DE DUCHER (H.P.)

The large globular buds of cerise-red subtly open to a rich magenta, large and reflexing. The blooms are usually carried singly on stout upright stems (1865).   3b. d. R.

GRUSS AN TEPLITZ (H.T.)

This Hybrid Tea features loose clusters of nodding, semi-double blooms of light red, deepening upon opening, and strongly fragrant. The growth is thin and twiggy, yet tidy (1897).   3d. sd. P.

LADY HILLINGDON (Tea)

One of the best known Tea-roses and a very early flowerer, it has long pointed buds opening to loose blooms of apricot-yellow on thin arching stems. The stem and foliage effect is one of attractive bronze-green (1910). The climbing form is scarcely more than shrub vigour but is equally attractive as the bush, especially in early spring.   Cl. or 2d. sd. P.

LADY MARY FITZWILLIAM (H.T.)

A very prominent breeding parent (1882), having influenced the majority of our present H.T.'s. For a while the variety was thought to have been lost, but it was recently discovered and re-introduced by Mr G. S. Thomas of England. The large globular blooms of rose-pink are borne on a stout bush. It is especially free flowering and fragrant.   3b. d. P.

LA FRANCE (H.T.)

This rose is generally considered to be the first Hybrid Tea (1869) when Mme Verdier, a Hybrid Perpetual, was crossed with Mme Bravy, a Tea. It has large, well shaped blooms, that have a tendency to ball, and are silvery-pink, perpetual flowering, and strongly fragrant. However, as the strain seems to have declined it is not recommended for other than historical interest.   2c. d. P.

MME ABEL CHATENAY (H.T.)

Surely the oldest rose to have remained in popular demand (1895), the blooms are of "modern" form with high centres and reflexing petals of shell pink with silver reverse, strongly tea-scented. The large open bush resents hard pruning, but is still an excellent rose by both old and modern standards. As a climber it is slow to develop, but eventually makes a large plant which blooms very freely.   Cl. or 3c. d. P.

MME VICTOR VERDIER (H.P.)

The stout upright stems and large matt foliage are typical H.P., with light red blooms tending silver on the reverse. From a globular bud the flowers reflex to a large loose ruffle. 3b. d. R.

MRS JOHN LAING (H.P.)

The large double blooms of almost "modern" shape are pink, shading deeper upon opening, and is richly fragrant (1887).   3b. d. P.

PAUL NEYRON (H.P.)

This rose has very large blooms of rose-pink, double and quartered, carried singly on stout upright stems, and is strongly scented. It has often been loosely called the Cabbage Rose (1869).   4b. d. P.

PRINCE CAMILLE DE ROHAN (LA ROSIÈRE) (H.P.)

A rather "modern" rose for its day (1861), with rich crimson blooms of reflexing form and richly scented. The growth is of usual H.P. habit, tall and upright.   3b. d. R.

REINE DES VIOLETTES (GENERAL STEFANIK is either synonymous or very similar) (H.P.)

This excellent rose could well be classed as a Bourbon, as the fragrant flowers are double and cupped with tight quartered centre, while the colour is violet-purple with silver-purple reverse. The long upright canes which carry small clusters of bloom are thornless. In all, highly recommended (1860). 3b. d. R.

ROGER LAMBELIN (H.P.)
A novel but striking rose, where the rather irregular crimson petals are sharply edged and divided white. The blooms are semi-double and slightly recurrent (1890).   2b. sd. R.

SOLEIL D'OR (Originally PERNETIANA but now classed as H.T.)
The second generation of *foetida persiana* × Antoine Ducher, this rose was the break-through which introduced the bright yellow and copper colours into modern breeding (1900). The light green foliage and loose globular blooms resemble *foetida*, although the colour is soft coppery-gold and the bush is more compact. It is prone to black spots in some climates. 2c. sd. P.

VICK'S CAPRICE (H.P.)
This 1897 hybrid is characteristically "old world" almost to the point of coarseness, yet attractive at times. Globular buds open cup-shaped with a tight quartered centre, revealing pale lilac petals softly striped deeper pink to magenta and strongly perfumed, on a clean almost thornless bush.   3b. d. R.

LANDSCAPING WITH SHRUB ROSES

LANDSCAPING by its very nature looks at plants as a whole—and their relationship to other plants—their size and outline, their colour and texture, the massed effect of the flowers, and only lastly, their individual blooms. This is in contrast to the trend of modern breeders, where the perfection of the individual bloom is paramount. It is important therefore to consider your landscape as a whole, into which you wish to plant shrub roses. The neat formal garden, the small contemporary setting, or the natural bushland area, all can do justice to a collection of shrub roses, given a sympathetic understanding of their styles and effect.

You will find that species shrubs give the greatest variety of effects. Thick low plants such as *rugosa* and *chinensis* make wonderful borders and foregrounds. Taller thicketing plants such as the Hybrid Musks, *fedtschenkoana*, *woodsii*, and the "Frühlings" types suggest dividing hedges, dense backgrounds, and large specimen plants. The tall but open plants such as *moyesii, sericea pteracantha, hugonis, soulieana, dupontii, willmottiae*, and *macrantha* hybrids are best used as informal semi-wild plants, allowed to intertwine and blend gradually into the bushland. The species climbers also lend themselves to informal planting, growing over fences and trees, hiding old sheds, and tumbling from supporting posts and tripods. Banksias, *laevigata*, Wedding Day, Mermaid, and *brunonii* are some that can be recommended. Of course we must not forget the prostrate growers such as *wichuraiana, paulii, paulii rosea*, Max Graf, and *bracteata* for covering banks, walls, rock outcrops, and similar situations.

For a focal point of colour throughout the year, the choice is wide. Many of the China roses, lower Tea roses, Hybrid Teas, and Rugosas provide a variety of colours the year round as a foreground planting. Taller colourful plants are found in Noisettes, Bourbons, and some Hybrid Perpetuals, and of course the Hybrid Musks and other more recent shrubs. Colour from the foliage and hips of species should not be overlooked. The red foliage of *rubrifolia*, the grey of *fedtschenkoana*, the red thorns of *sericea pteracantha*, and the hips of *moyesii*, *dupontii*, and *rugosa* are but a few.

A ROW OF INTERESTING ROSA SPECIES
for foliage, thorns and hips

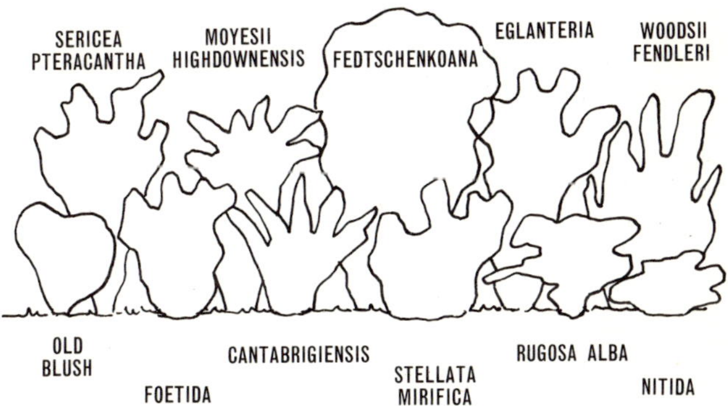

The "Old European" group of Gallicas, Albas, Damasks, Mosses, and Centifolias owe most of their charm to their flowers and fragrance in spring. For this reason they should be planted in a bed of their own which comes to the fore in late October and November, a site not hidden from view but nestled conveniently with your other spring shrubs. It is here that a word must be said in defence of spring flowering roses. Modern perpetual flowering roses have thoroughly spoiled us, for the tireless efforts of the rose breeder have given us a genus that is unsurpassed for continuity of flower. Many other plants hold a popular place in our gardens by virtue of an annual concentrated burst of colour, either as flowers or foliage. And yet the earlier roses, with a similar burst of colour, are despised because they cannot match the performance of their modern brethren.

To test the point, make an unbiased comparison between a spring flowering rose in full bloom and one of your favourite garden plants, either native or exotic, and see whether the rose does not acquit itself surprisingly well. Then consider the rose's perpetual flowering habit as an additional bonus to an otherwise remarkable plant.

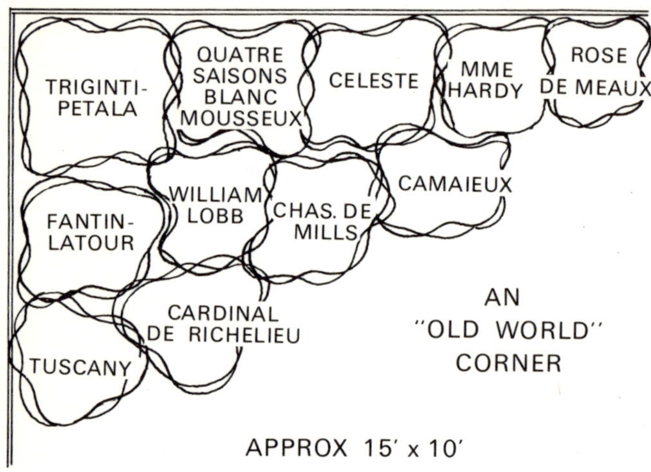

I cannot offer you any strict formula for designing your rose beds. The species suggest an informal layout with freestyle curves and mixed plantings. The "Old European" group suggest a more formal, closely planted bed to form a thicketing mass of well-graded colour. As the classes become more modern, they tend more towards the straight line symmetrical layout. Do not make the beds so deep that the centre plants are inaccessible. Similarly, make sure that the spacing of your front rows are close enough to make a continuous effect, spacing the back rows wider where the spaces will not be noticed, thus providing easier access. A transition from shrub roses to moderns can be achieved smoothly through Chinas and Tea roses, while for the transition into a bushland setting I would suggest the use of the open woody species such as *R. foetida*, *primula*, *eglanteria*, and *woodsii*. I have intentionally stressed the use of roses in beds or rows on their own because of the more concentrated and united effect that they give. In addition, their maintenance and culture is so much easier.

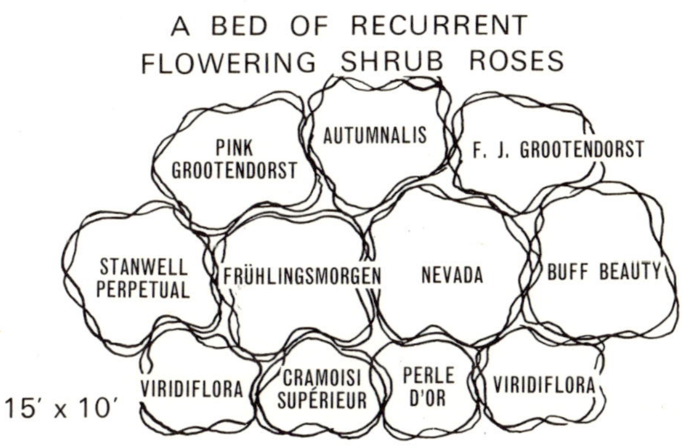

The accompanying sketches should give you a starting point from which to work. They may be used in their entirety, or they may provide a guide for your own planning, bearing in mind their relative sizes, habits of flowering, and blends of colours. The varieties suggested are among the best of their particular groups' consistent with a wide range of colours. However some excellent bedding, hedge, and climbing varieties that do not appear in the sketches are listed as "Recommended Shrubs."

Much of the atmosphere of your shrub rose garden will depend upon the minor details. Underfoot the feeling should be restful and soft. Turf that gently follows the outline of each plant and invites you from one plot to the next, a deep layer of leaves, chips or bark underfoot, and slices of tree-trunk laid as steps through the closer plantings, are all effective. The old tree-trunk or rocky outcrop will add to the atmosphere, while small voluntary clumps of daffodils or other winter bulbs look the part. A bird bath or birdhouse not only looks friendly, but encourages those busy little insect eaters into the garden. Carry the natural unsophisticated treatment into your fencing and woodwork with rough hewn or sawn textures, perhaps stained to your liking. Label your plants by all means, but use subdued colours that blend into the setting, with lettering of suitable size to be read when standing beside the plants.

A GARDEN OF RECURRENT 'OLD WORLD' ROSES
APPROX 20′ x 20′

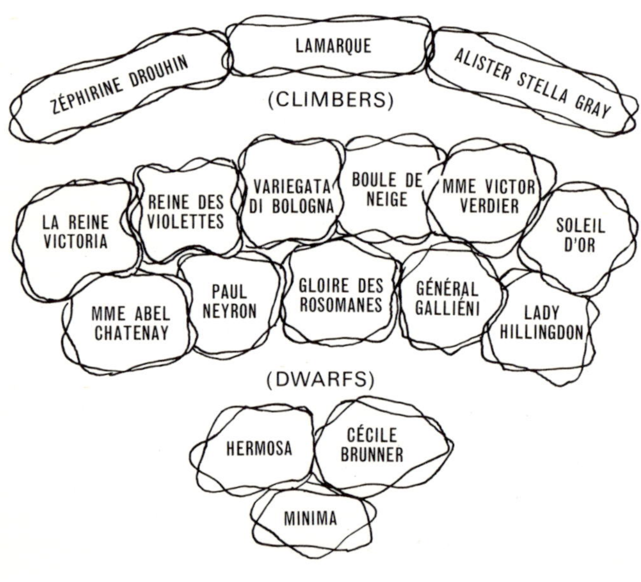

FRAGRANCE

WITHOUT doubt, much of the charm and popularity of the rose depends upon its fragrance, or more precisely, the subtly different perfumes that characterise the various groups. To this point most rose lovers will agree, but on their personal perception of each rose they will disagree. Rose fragrance is a very elusive factor. It waxes and wanes with the time of day, temperature, and humidity. It varies from person to person, and varies sometimes from day to day for each person. It can be said that all roses have some scent for some people at some time of the season. Beyond this it is unwise to be dogmatic, as everyone must accept the rose's fragrance as they find it. References to fragrance throughout this book are based on the strongest, most popularly accepted fragrances. It will not be surprising if you disagree with at least some of the descriptions, for such is the elusive rose fragrance.

But this I do know—on a warm spring evening as the sun sets, I can walk through the rose garden, catching the refreshing atmosphere as I pass each group. The pungent smell of the *foetida*, the sweet refreshing *eglanteria*, the rich Musk row, and the almost overwhelming Bourbons and Hybrid Perpetuals. I do not sample individual blooms, but sense the fragrance of the whole bush at a distance—a truly enriching experience.

CULTURE

Shrub roses, and especially those more closely approaching the wild species, are particularly hardy. Like all roses they prefer an open sunny position in an area of their own, away from the robbing roots of trees, shrubs, and lawns. However most will tolerate difficult conditions more readily than the moderns. Indeed I have seen them growing remarkably well when confined above, below, and around by huge shrubs, but I would not recommend that you push your luck that far! Like other roses, it is important to keep the soil in good condition with a steady application of mulch in the form of old leaves, grass, straw, seaweed, bark, woodchips, or peat. Digging of the soil is unnecessary when the mulch is working effectively. Animal manures included with the mulch are a great help, while a winter dressing of a "complete" rose manure is recommended. Avoid excessive nitrogen fertilising, such as blood and bone manure, as this gives soft, disease prone growth. Water when necessary throughout the year, with heavy soakings every two or three weeks rather than surface sprinklings.

Shrub roses possess more than the average resistance to disease. Powdery mildew is perhaps the most troublesome, affecting at times the "Old Europeans," *laevigata* and its hybrids, some *wichuraiana* hybrids, *hardii*, and some Chinas. Sulphur dusting in the warm weather is still a good natural treatment. Rose Rust may occur on some Hybrid Perpetuals, and is treated by spraying with Zineb. Blackspot does not seem to be a major problem, appearing mainly on some yellow varieties. Insect pests in the form of aphids, thrips, tortryx grub, and curculio beetle do occur, but usually less severely than on the moderns. The encouragement of birds and insects, together with hardy, non-forced growth, is the best and most natural control. When these methods have failed, use only the specific spray for the pest in question, as many "do-all" sprays actually do nothing properly. I have great faith in a thorough winter spray when the plants are dormant. A combined copper (Bordeaux or copper oxy chloride) and tar distillate (Winter Oil) spray will control the over-wintering spores and eggs, thus making the spring and summer control far easier.

Pruning recommendations have been given briefly in the introduction to each group. Broadly speaking, the perpetual flowering varieties are pruned in winter while the spring flowering ones are pruned after flowering. However, be guided by the manner in which the variety bears its blooms. Those that flower from short stems off the old wood, such as Banksias, *moyesii*, and *hugonis*, would lose much of their potential bloom if pruned in winter. On the other hand, others like the "Old European" group for example, make relatively long stems after they have finished flowering, and these may be lightly shortened in winter.

In most cases pruning should never be heavy. Remove wood that is obviously smothered or worn out, but only lightly trim the new canes. Long canes may be trained outwards and twined around neighbouring canes, or arched over the top of the bush. Climbers on tripods, posts, or trees should be trained in spiral fashion to keep the canes as horizontal as possible.

FLORAL ARRANGEMENTS
AND COMPETITIONS

Sooner or later you will feel compelled to cut a bunch of your shrub roses for interior decoration. It may be for your home, or it may be for a show, but in either case, they can make a very attractive display, although they are somewhat different to the moderns in treatment and effect. For example, the muted colours of the old-world types, the whites to pinks to dusky reds to purples all blend without a jarring tone, so that even the most casual arrangement looks well blended.

Arrangements should be massed, rather than trying to emphasize individual blooms, using the well known guidelines of "deep colours at the centre, light colours to the extremes," and "larger, open blooms at the centre, smaller blooms and buds to soften the extremities." The nodding, arching habit of many varieties makes it easy to use a cascading effect to advantage. Containers should be of period style such as ornate silver, thick crystal, bronze, pewter, or marble.

The simpler species forms suggest a more stark, angular effect, using whole stems of bloom and hips if possible, with a background of fine young foliage. Containers may be of pottery, copper, wood, or other natural texture, with accessories of driftwood, bamboo, pebbles, or terracotta.

It is important to condition your blooms before use if you expect them to keep for more than a couple of hours. Cut the blooms in the cool of the day, plunge them immediately into deep tepid water and leave them literally up to their necks in water in a dark position for several hours. Splitting, scraping, of obliquely cutting the stems provides a larger drinking surface and is worth trying. Pinpoint holders usually split the stem sufficiently to also help. Start with clean containers, use clean water and add a drop of mild disinfectant to keep the water from becoming slimy. Singles keep better if cut as a bud and allowed to open indoors; in fact, large singles such as "Scabrosa" will last several days when prepared in this way.

The major Rose Shows in Australia usually provide a class for "Old-fashioned Roses" but it is usually small and unimaginative, merely calling for "three cuts" or sometimes including "Singles—three cuts" or "Species—three cuts." Show schedules often do not define an "Old-fashioned" rose, and competitors are left in doubt as to the standards by which their entry will be judged. I prefer to classify an "old-fashioned" cut by the form of the flower and not by the date of its introduction. For instance, Constance Spry (1961) would be eligible for the class but Maréchal Niel (1864) is of almost modern form and therefore more appropriate with the "garden" section. The "old-fashioned" cuts should be staged as open blooms with the stamens fresh, or the ruffled or button-eyed centres crisp and distinct. They should be judged upon their circular outline, symmetry of form and development, freshness, and freedom from damage. Cuts should be staged with stems as long as possible, and so arranged that the blooms all squarely face the judge in distinct rising stages towards the back row.

Singles, because of their fragile nature, are more difficult to stage. Where possible, cut a branch with plenty of buds that will continue opening when cut. Take spare branches in case one "falls" completely, conditioning the cuts as you would for floral arrangements. Singles, of course, are staged fully open but must be fresh and free of damage; the condition of the stamens is a good indicator of freshness. Although a circular outline is desirable, it should not be binding, as some singles never achieve this and yet make excellent exhibits.

"Old-fashioned" and singles should not be disbudded, but unsightly spent blooms may be neatly and inconspicuously removed. Finally, name your entries as correctly and legibly as possible. It not only helps the Show officials but it may also awaken someone's interest in shrub roses.

RECOMMENDED SHRUBS AT A GLANCE

Gallicas, Damasks and Albas
alba maxima (A), Assemblage des Beautés (G), Belle Amour (A-D), Cardinal de Richelieu (G), Celeste (A), Duchesse de Montebello (G), Königin von Dänemark (A), Mme Hardy (D), Président de Sèze (G), Sissinghurst Castle (G), Tricolore de Flandre (G), Trigintipetala (D), Tuscany (G).

Centifolias and Mosses
centifolia (C), Chapeau de Napoléon (C), *centifolia muscosa* (Common Moss), *centifolia muscosa alba* (M), Comtesse de Murinais (M), Fantin-Latour (C), Japonica (M), Nuits de Young (M), William Lobb (M).

Noisettes and Bourbons
Alister Stella Gray (N), Jaune Desprez (N), Lamarque (N), La Reine Victoria (B), Maréchal Niel (N), Mme Isaac Pereire (B), Souvenir de la Malmaison (B), Variegata di Bologna (B).

Hybrid Perpetuals, Tea and Hybrid Teas
Baron Girod de l'Ain (HP), Lady Hillingdon (T), Devoniensis (T), Général Gailiéni (T), Mme Abel Chatenay (HT), Mme Victor Verdier (HP), Paul Neyron (HP), Reine des Violettes (HP).

Chinas
Bloomfield Abundance, Cécile Brunner, Hermosa, Minima, Old Blush.

Species and Close Hybrids
cantabrigiensis, Carmenetta, *ecae*, *fedtschenkoana*, Frau Dagmar Hastrup, *highdownensis* (*moyesii*), Mutabilis, *pomifera duplex*, *roxburghii plena*, *rugosa alba*, Scabrosa, Stanwell Perpetual.

Climbers
Albertine, Banksias, Debutante, *laevigata*, Lamarque, Mermaid, Sea Foam, Wedding Day.

Ground Covers
bracteata, Max Graf, *paulii*, and *paulii rosea*, Sea Foam, *wichuraiana*, and some hybrids.

Modern Shrubs
Autumn Delight, Buff Beauty, Felicia, Frühlingsmorgen, Iceberg, Nevada, Scarlet Queen Elizabeth.

Low Foreground and Edging Plants
Several China varieties, Frau Dagmar Hastrup, *nitida*, *Raubritter*, Rose de Meaux, William III.

Good Specimen Plants
Constance Spry, *fedtschenkoana*, Iceberg, Mme Plantier; most Hybrid Musks, especially Buff Beauty, Felicia, Bonn; Nevada, and Marguerite Hilling; Stanwell Perpetual; most Rugosa, especially Scabrosa.

Strong Impenetrable Hedges
fedtschenkoana, *hugonis*, *macrantha*, *micrugosa*, *multiflora*, Schneelicht, Trigintipetala.

Some Thornless or Almost Thornless Shrubs
Banksias, fortuneana, Mme Sancy de Parabère, Nevada, Tausendschön, Veilchenblau, Zéphirine Drouhin; Chloris, Mme Legras de St Germain, Mme Plantier (Albas); Georg Arends, Mrs John Laing, Paul Neyron, Reine des Violettes (Hybrid Perpetuals).

Some Shrubs and Climbers that Set Spectacular Hips
Shrubs: *moyesii* types (except Eos); Rugosas *alba* and *rubra*, Scabrosa, *calocarpa*, Schneezwerg; *alba maxima*, *dupontii*, Frühlingsmorgen, *pomifera duplex*, *macrantha*, *rubrifolia*, *roxburghii plena*, *setipoda*, *spinosissima altaica*, *stellata mirifica*, *woodsii fendleri*. Climbers: *brunonii*, *helenae*, Wedding Day.

Striped and Parti-coloured Blooms
Camaieux, Commandant Beaurepaire, *damascena versicolor* (York and Lancaster), *gallica versicolor* (Rosa Mundi), Ferdinand Pichard, Georges Vibert, Honorine de Brabant, Tricolore de Flandre, Variegata di Bologna.

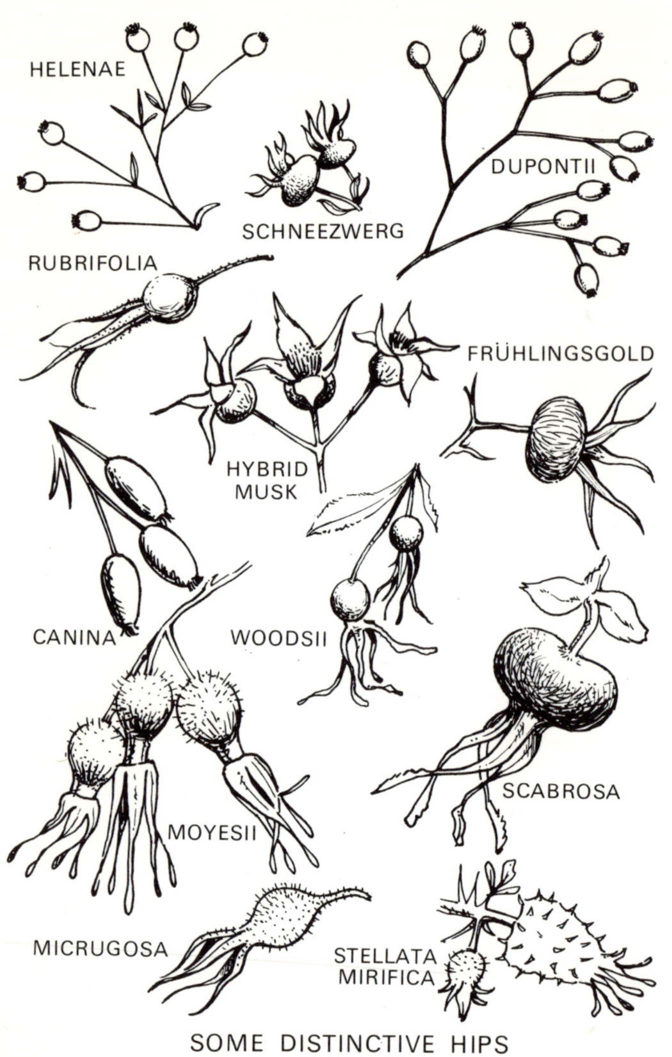

HELENAE
SCHNEEZWERG
DUPONTII
RUBRIFOLIA
FRÜHLINGSGOLD
HYBRID
MUSK
CANINA
WOODSII
SCABROSA
MOYESII
MICRUGOSA
STELLATA
MIRIFICA
SOME DISTINCTIVE HIPS

Shades of Blue, Grey and Magenta

Belle de Crécy, Bleu Magenta, Cardinal de Richelieu, Hippolyte, Jenny Duval, Nuits de Young, Président de Sèze, Reine des Violettes, Tour de Malakoff, Tuscany, Veilchenblau.

Varieties which were the original of their class or were important parents, and are therefore of BOTANIC or HISTORIC interest.

chinensis semperflorens (Slater's Crimson), Old Blush (Parsons' Pink), *chinensis minima* (Rouletti), *moschata, wichuraiana, multiflora, foetida persiana*; DAMASKS: *trigintipetala, versicolor, bifera*; ALBAS: *maxima*, Maiden's Blush; GALLICAS: *officinalis, versicolor*; CENTIFOLIAS: *centifolia, cristata, muscosa, muscosa alba;* BOURBONS: Rose Edouard, Reine de Isles Bourbon (Bourbon Queen); NOISETTES: Champney's Pink Cluster; Portland Rose; HYBRID TEAS: La France, Soleil d'Or, Lady Mary Fitzwilliam.

Shrubs with Curious or Special Features

Bullata (bullate leaves), Chapeau de Napoléon (crested "Moss"), Japonica (heaviest moss), Viridiflora (green "flowers"), *sericea pteracantha* (thorns), *ecae* and *moyesii* (brilliant colours), Mutabilis (changing colour), *rubrifolia* (foliage); Fimbriata, Pink Grootendorst, F. J. Grootendorst, *serratipetala*, (frilled petals); *primula, eglanteria*, Golden Moss (aromatic foliage); *hardii, roxburghii, stellata*, Banksias ("borderline" species).

A Short List

I am often asked to state my favourite rose. Frankly, it is easier to choose ten or one hundred than it is to choose one. Those listed below are some of my favourites in the various groups, and could well be a starting point for your collection.

Buff Beauty, *cantabrigiensis*, Cécile Brunner, Chapeau de Napoléon, Constance Spry, Hermosa, La Reine Victoria, Mme Hardy, Mermaid, Nevada, Scabrosa, Sea Foam, *sericea pteracantha*.

INDEX

Duchess of Portland, *see* Portland Rose
dupontii, 6
Düsterlohe, 8

ecae, 6
Echo, 30
eglanteria, 6-7
Eglanteria Lady Penzance, *see* Lady Penzance
Eglantine Rose, *see* eglanteria
Elmshorn, 21
Empress Josephine, *see* Francofurtana
Eos, *see* moyesii eos
Eva, 21

Fairy Rose, *see* minima
Fantin-Latour, 46
farreri persetosa, 7
Father Hugo Rose, *see* hugonis
fedtschenkoana, 7
Felicia, 21
Félicité Bohain, 46
Félicité et Perpétue, 15
Félicité Parmentier, 38
Ferdinand Pichard, 56
Filipes "Kiftsgate", 15
Fimbriata, 26
F. J. Grootendorst, 26
foetida bicolor, 7
foetida lutea, 7
foetida persiana, 7
fortuneana, 15
Fortune's Double Yellow, 15
Francofurtana, 38
Frankfort Rose, *see* Francofurtana
Frau Dagmar Hastrup, 26
Frau Karl Druschki, 56
Fritz Nobis, 21
Frühlingsanfang, 21
Frühlingsgold, 22
Frühlingsmorgen, 22
gallica versicolor, 38-39

Général Galliéni, 56-57
Général Jacqueminot, 57
General Kléber, 46
General Stefanik, *see* Reine des Violettes
Georg Arends, 57
Georges Vibert, 39
Geranium, *see* moyesii geranium
gigantea, 15
Gipsy Boy, *see* Zigeunerknabe
Gloire de Ducher, 57
Gloire des Rosomanes, 30
Golden Moss, 46
Golden Rambler, *see* Alister Stella Gray
Golden Wings, 22
Goldfinch, 15
Gold of Ophir, *see* Fortune's Double Yellow
Great Double White, *see* alba maxima
Green Rose, *see* viridiflora
Gruss an Teplitz, 57

hardii, 8
harisonii, 8
Harison's Yellow Rose, *see* harisonii
Heather Muir, 11
Hebe's Lip, 39
helenae, 15
Henri Martin, 46
Hermosa, 30
highdownensis, *see* moyesii highdownensis
Himalayan Musk Rose, *see* brunonii
Hippolyte, 39
Honorine de Brabant, 52
hugonis, 8

Iceberg, 22
Incense Rose, *see* primula
Ispahan, 39

Assemblage des Beautés